COLLECTING SHOTGUN CARTRIDGES

COLLECTING SHOTGUN CARTRIDGES

Ken Rutterford

Stanley Paul
London   Melbourne   Auckland   Johannesburg

Stanley Paul & Co. Ltd

An imprint of Century Hutchinson Ltd

Brookmount House, 62–65 Chandos Place,
Covent Garden, London WC2N 4NW

Century Hutchinson Australia (Pty) Ltd
PO Box 496, 16–22 Church Street, Hawthorn, Victoria 3122

Century Hutchinson New Zealand Limited
PO Box 40–086, 32–34 View Road, Glenfield, Auckland 10

Century Hutchinson South Africa (Pty) Ltd
PO Box 337, Bergvlei 2012, South Africa

First published 1987
Copyright © Ken Rutterford 1987

Set in 10 on 11pt Ehrhardt Roman
by Input Typesetting Ltd, London

Printed and bound in Great Britain by
Butler and Tanner Ltd, Frome

All rights reserved

British Library Cataloguing in Publication Data

Rutterford, Ken
   Collection shotgun cartridges.
   1. Cartridges — Collectors and collecting
   I. Title
683.4′06   NK6904

ISBN 0 09 166330 X

Contents

Acknowledgements	6
Foreword	7
Introduction	8
Building a Collection	9
Where to Obtain your Cartridges	10
Displaying a Collection	11
Refurbishing and Loading Tools	13
Other Collectables	17
Pinfire Cartridges and Others	18
How Old is It?	22
The Cartridge List	24
Cartridge Code	25
Firms of the British Isles and their Cartridges	32
Some Unidentified Cartridges	123
An Australian Collection	124
Australian Cartridges Post Second World War	129
Names of Firms to be found within the Cartridge	130
Headstamp Letter Combinations on Shotgun Cartridges	132
Significant Events	135
Index	137

Acknowledgements

I EXTEND my warmest thanks to the many members of the British Cartridge Collectors Club who have so kindly provided me with information on firms long since vanished and the cartridges they produced.

I would also like to thank the numerous people who have stopped to talk to me about the old days of cartridge loading when I have visited ironmongers' and gunsmiths' shops. All this information has been of enormous value. I am indebted to other enthusiasts who have allowed me to inspect their collections enabling me to record cartridges which I do not have in my own.

Some of the collectors who have been of great assistance are Gerald Barrell, Colin Caws, Derek Cullum, David Fleet, Chris Goodchild, Frank Gordon, Peter McGowan, Gordon Parker, Ciscel Twyman, Eric Wastie, Martin Woolhead, Alan Yeates and Bill Young. There are many others too numerous to mention but I am grateful to them all.

In particular I must thank Ronnie Crowe for the splendid colour photographs which illustrate the book and last, but by no means least, my family for putting up with the continuous tapping from my typewriter.

Foreword

THERE IS rather more to a shotgun cartridge than one might at first imagine, each one in its own way is a small piece of history telling the enthusiast something about the firm which produced it but also giving an insight into the weapons and the sport enjoyed at the time. This splendid book by Ken Rutterford will surely become the standard reference work for the growing band of enthusiasts of this fascinating subject. The advice to collectors is comprehensive, the technical aspects are dealt with accurately and in detail and the recommendations on refurbishing old cartridges will be welcomed by the expert and beginner alike.

But it is his compendium of individual cartridge types where Ken Rutterford's vast knowledge and experience comes into its own. Before the cartridge industry as we know it today came into being, unloaded cases were often supplied from the manufacturers, Eley, Kynoch etc to be loaded by gunsmiths and ironmongers to their own specifications. There were literally hundreds of different types of cartridges made. According to personal taste sportsmen could choose from a vast range of cases made with paper tube, metal covered paper, waterproof paper (the famous Eley patented 'Pegamoid' process), coiled metal or solid drawn brass.

The diversity of these different types was a nightmare for the cartridge loaders of the day who, under the guidance of the industry, later standardised on the popular components which are still in use today. Any frustrated creativity was spent in inventing the most colourful of brand names. What sportsman would not be drawn to a cartridge called 'Hobson's Full Stop' but there are many others to choose from among the hundreds reviewed here.

Ken Rutterford's work will be a constant companion for the growing number of enthusiasts caught up in the fascination of cartridge collecting.

J L Saint
Marketing Operations Manager
Eley Limited

Introduction

THIS BOOK is not intended to be a history of the major British cartridge companies such as Eley Brothers, Kynoch Ltd, Frederick Joyce or Nobels. Much material has already been published on them. My chief reason for producing this work is to introduce the fascinating hobby of shotgun cartridge collecting and to place on record, as far as I am able, the history of these old cartridges relating to the British Isles.

I believe most people have a touch of the magpie in them. Some collect old bottles, coins or stamps all their lives but I confess that my addiction to cartridges came later in life. I was doubtless born with the acquisitive bug but it lay dormant for years and then, when it bit, I became really hooked.

The son of a gamekeeper, I was born well off the beaten track. We lived in a keeper's cottage in Hampshire woodland in the parish of Amport. The trains that ran over the Grateley Incline and the wildlife were the only signs of activity in the area. Each season between 1928 and 1934 the guns would come to Sarson Wood and stand in front of our house. Often Royalty was amongst them and when they left the stands would be littered with empty cartridge cases.

After the firing was over and the guns had departed, I would collect some of the colourful cases. My mother used to claim that at Christmas I would discard my new toys in favour of my box of empty cartridge cases!

As a schoolboy I built up a small collection of cartridge cases, but somehow they vanished, some perhaps finding their way back to the gunsmith for reloading. During the Second World War cartridges became difficult to obtain and good cases were welcomed by those who had reloading equipment.

Much later in life the collecting mania gripped me. Whilst looking through a magazine I came across an advertisement from a gamekeeper seeking old cartridges for his collection. I noted his address and paid him a visit. He was a Mr Gordon Parker, a partridge keeper living on the plain at Upavon. His collection fired me with enthusiasm and though not greatly struck by the few cases I had brought he was kind enough to give me some old ones to start me off. Since then I have never looked back.

A fine collection can be a reference and a history of ironmongers, gunsmiths and gunmakers and their addresses over many years. To those of you who take up this challenge, I wish you well. I am positive you will make new friends and spend many happy hours absorbed in this interesting hobby.

Building a Collection

ALTHOUGH my collection contains many old cartridges from the beginning of the introduction of the breech-loading system, someone starting a collection today will have difficulty in finding many of the early pinfire and centre-fire cartridges. However, the occasional sample still turns up and in even the smallest collection there will often be one or two gems.

Do not be disheartened by the fact that few of the old rolled-top cartridges are around. New brands of cartridges are coming on to the market nearly every week and many of these are just as colourful and decorative as were their predecessors. By joining collectors' clubs the door is opened to collectors throughout the world. Remember, though, that it is forbidden to send live ammunition through the post. Only fired cases, uncapped cases or old cartridge boxes are allowed.

I dislike unloading or firing old cartridges. To me this is a sacrilege. When they have been found and are in good condition then keep them that way. In the past I have been guilty of ruining several good old cartridges while removing the powder charge and trying to refill them with a harmless substitute. There can be some danger in pressing out live caps so take care. I have pressed out many when stripping down old cartridges that are beyond saving so that I can retain the caps for use in other salvageable cartridges. If you decide to fire a ready capped case make sure there is no live charge hidden behind the wad in the bottom. If there is, a corkscrew will remove it. Remember that explosives can bite back so be ultra cautious at all times.

Old primer caps are very corrosive so never fire them in a good gun. Prepare the barrel of the gun by pushing a small piece of oil-soaked rag through it. This will leave a thin coat of oil on the interior. Now wipe out all the oil from the chamber to a depth just below that of the cartridge case. In order to avoid any damage to the case, test a similar case that is not required. Keep the barrel of the gun pointing down so that no oil runs back into the chamber.

When firing the cap keep the gun pointed out of harm's way as there is a surprising amount of power in an old cap. Never fire 16-bore cases in a 12-bore except solely for the purpose of firing the cap, and then you can push a paper sleeve cut from a 12-bore case over the 16-bore case. This has the added advantage of keeping the case from any possible oil in the chamber. As soon as you have finished detonating percussion caps clean the gun. Later will not do.

Should you be lucky enough to obtain some vintage cartridges it would be unwise to shoot them off. If the paper tubes show any signs of damp or swelling you will be asking for trouble. I have stripped down old black powder cartridges that have been subjected to damp and have discovered that the powder charge has solidified.

Where to Obtain your Cartridges

MANY of my older cartridges have been obtained from gamekeepers and farmers. A farmer who has everything in his yard from a horserake to a combine harvester, all in a state of decay and rust, is just the sort of chap to have a few old cartridges tucked away in a box or drawer. You may be lucky right away or it may take some time before they are unearthed. Often cartridges have been hidden away from the light of day for a great many years in the back of an old cupboard or chest of drawers.

Gunshops and old ironmongers that have not been modernized are always worth a try but most of the old cartridges have been thrown away or found their way into collections. However, one or two trickle into these shops from time to time and it pays to keep on friendly terms with your local gunsmith.

Modern cases are more easily acquired. Shooting folk will save a few of each cartridge that they shoot or come across during their travels. As cartridges are expensive, some reimbursement should be offered when someone takes the trouble to save them for you, even if they are only spent cases.

During the pheasant and grouse shooting seasons, a word or two with the local keepers, pickers-up or loaders can often yield a few cartridges. Never trespass in order to try to increase your collection. The majority of keepers are only too ready to help provided you go about it in the right way. I know of several keepers who are themselves keen collectors.

Clay shoots are another good source of cases. Wait until the shoot is over when your help in clearing the ground of fired cases will be appreciated.

Cartridges may turn up anywhere; an old cottage being cleared, arms sales, in fact in a hundred and one different places. Remember that it is always advisable to have an exchange box and to keep it well stocked up. From time to time you will be offered cartridges, but you must have something to offer in return. If you are given a good lead follow it up. Cartridges do not come to you, you have to seek them out.

Cartridge collecting is not restricted to the British Isles, though many collectors in this country prefer to specialize in the cartridges from Britain. Through this hobby I have made friends with many other collectors and their families. Without the cartridge the gun would be useless, so why not build yourself a collection of cartridges or cartridge cases? Cases are safer but they do not hold for me the same interest as the live round. I have many fired cases in my own collection, some of which are the last examples of their type. Should a live example of any of these turn up then it replaces the spent case.

Displaying your Collection

HAVING seen many collections I have been able to assemble the best ideas for displaying cartridges. I have seen collections kept in cardboard boxes and others housed in well-constructed cabinets. If you must, keep your cartridges in boxes, but if you want to display them and maintain their condition you will need to consider some form of cabinet. This becomes even more urgent as the collection grows.

One method is a wall cabinet. Here the cartridges stand in rows on narrow shelves and are often viewed through glass-fronted doors. Though this is eye-catching there are several pitfalls. Firstly, this type of cabinet will not hold nearly as many cartridges as one with drawers, whilst if live cartridges are kept their sheer weight can become a problem. Over a period of time the shelving will start to sag. In addition, they become subject to infra-red rays from sunlight and the colours on the paper tubes quickly fade. Pink, blue and green tend to fade first. One collector, although he has his display cabinet in a dark passageway, has had to protect the cases for this reason. He has had the happy idea of pasting labels from old cartridge boxes over the glass.

The best way to house a collection is in a cabinet with drawers. It must be strong as it will have to bear a considerable weight. The drawers or trays should be shallow and in these the cartridges are laid in neat rows. As the drawers are opened, the printing should face one way so that you can read it. Wherever possible the cabinet should have closing doors which can be locked.

In the past I have converted cabinets to suit my own purpose and if money is no object a cabinet can be purpose built. Cabinets that have been built to house maps or cotton reels, for instance, may require strengthening. Metal filing cabinets with rows of small drawers are very suitable. They have the advantage of being fireproof and often have a device which will lock all the drawers.

To place cartridges in straight rows in the drawers can be difficult. As the drawer slides in and out they will roll about. If the drawers are wooden, then thin strips of wood may be glued across to keep the cases in rows. These strips also act as strengtheners. Model aircraft wooden stringers are useful but expensive. Some DIY shops sell thin beading, which is much cheaper. If you have a power drill then boxwood stringers can be cut and, with some sanding, will be adequate. One idea I have found to be successful is to take 0·015-inch white card sheeting and to fold it at half-inch intervals as one would fold a paper fan. This, once folded, is then cut to the size of the cabinet drawers and inserted with the cartridges resting in the rows of channels. Many cartridge collectors in Britain know Mr Wastie. He uses corrugated cardboard in his cabinet drawers. Now there is nothing new in this except that he has improved the system by flattening alternate corrugations, depending on the size of the corrugated card and the gauge of the cartridge. His simple method is most successful and well worth emulating.

I keep the cartridges in my collection in alphabetical order by firms' names. This is useful if you wish to locate a cartridge in a hurry. Make sure that the glue you use does not have an acid hardener as this will attack the brass end of the cartridges and corrode them.

All live cartridges are best kept in a dry atmosphere, away from excessive heat and where air can freely circulate round them. Always keep them out of the reach of children and locked away.

An early loading machine for mass production (see also page 14)

Refurbishing and Loading Tools

THE MAJORITY of cartridges will need some attention before they can be placed in a cabinet if the appearance of the collection is to be maintained.

Old cartridges usually need to have their paper and brasswork cleaned up, while modern fired cases with crimp star closures can also be improved. I prefer to see all modern crimp closure cartridges dummy loaded before being placed in the collection. If this is not done then I suggest you improve their appearance by smoothing the end of the crimp using two pieces of wood.

I do not bother to replace fired caps but I do like to see the crimp closures turned back. To do this the case must be refilled with dry sand, which tends to be heavy, or a similar substance. A $^{1}/_{16}$-inch card wad is then inserted to retain the filling. Pushing back the crimp of paper cases by hand takes a little practice, but if you have a modern reloading tool then the problem is solved.

When using sand, the cases should be filled to just below the rim of the turnover. The height is determined by the type of case and the material used. Remember that some materials are more easily compressed than others.

Having chosen a material to fill the case you can now push back the star crimp by applying gentle pressure to the centre of the star using an object with a flat end and similar diameter. Push each point of the star downwards and inwards from the outer rim one at a time, working all round the top of the case. The crimp should now return quite easily and you will obtain a respectable dummy load.

If you prefer not to dummy load the next best thing is to smooth the crimped end to give the case the appearance of being unloaded. To do this, take a piece of wooden dowelling or metal bar the same diameter as the inside of the case. This should have a tapered lead to assist when inserting it into the case. Now roll the case to and fro between a flat piece of wood and a table top, keeping the rim of the cartridge away from the table and jutting over its edge. This should help flatten the crimp. Warming the case in front of a heater will often help soften the texture of the tube.

On the subject of crimp cartridges I only accept one of each kind in my collection, whereas some collectors take one of each shot size. This, I believe, is going too far and the collection soon starts to take over the house!

Since collecting cartridges I have been accumulating a selection of old hand-loading machines. Most of these tools are bench closing designs and are available in a variety of bores and though they are ancient they are well worth collecting. Designed at a time when cartridges were loaded by small firms throughout the length and breadth of the country, they brought the art of cartridge loading into the home. Many an evening has been spent

Note powder, shot, felt wads, cards, capped cased and 'four at a time' rammer

with the family sitting round the kitchen table passing cartridges from one to another as they went through the stages of loading. The first person would resize the case and pass it to the next, who would examine and recap it. The third would insert a measure of powder and insert and press home a wad. Number four would measure in the shot, ink the shot size on a $1/16$-inch white card wad, finally placing it on top of the shot. It was then handed to the fifth person, who would wax the case top and close it with a few brisk turns in a closing machine. As he or she finished it the cartridge would be inserted in a box or cartridge bag. Remember they had big families in those days! The recognized closure then was a rolled turnover, a form still used today for heavy loads with large shot.

If you come across any old loading machines they are well worth keeping, especially the closing machines. Mostly made of brass, they consisted of a chamber to hold the cartridge, at one end of which was a hand-operated lever which gripped the base of the cartridge, while at the other end was a turn handle. This latter revolved around a chuck which turned over the top of the paper tube so forming the closure. Most machines had a clamp with a thumbscrew so that it could be fixed to a workbench. Many of the early machines had a slot down the side of the chamber to enable pinfire cartridges to be loaded. Not all the tools gave the same results. Some chucks produced a round top to the roll, some a fat or square finish and on others the finish could be altered by turning a pin in the chuck.

To use a standard closing machine take a loaded cartridge with its shot load just about one quarter of an inch (7mm) below the top of the tube. Insert a $1/16$-inch card wad and then place the loaded round into the chamber of the machine. Close the machine with the handle so that the top of the case rests in the chuck. Apply a slight pressure on the lever while giving two or three sharp turns with the winding handle at the other end. This will roll over the top of the tube and close the load tightly below the wad.

It was always recommended that a little candle wax should be applied to the top of the case. I have found that a small piece of polythene of the type used for freezer bags placed between the chuck and the top of the cartridge tube will act as efficiently. A plastic case, when rolled, requires no waxing.

If plastic cases which have been fired are collected, they look far more attractive if they are dummy loaded. After filling and sealing with a wad, they can have their star crimps pushed back by hand. Hold each case to the front of a heater to soften the plastic and at once insert the cartridge into the closing machine and give it a very gentle turn. This will reset the rolled edge on the star crimp.

Older cartridges may require additional attention. Brass that is corroding will need to be cleaned, but do not use a metal polish as this can cause further corrosion at a later date. Instead dab at it gently with fine wet and dry paper and then finish off with fine wire wool. Remember that when you rub a surface you create heat through friction, something to which no cartridge should be subjected. Cleaning must be a slow, painstaking task. Once the corrosion has been removed and the brass polished enough to remove all scratch marks, then it can be treated.

Mask off the paper tube from the metal with a tight band of paper and with a pressurized container of polyurethane clear, hard lacquer, lightly spray a film over the metal base and then put it aside to dry for a couple

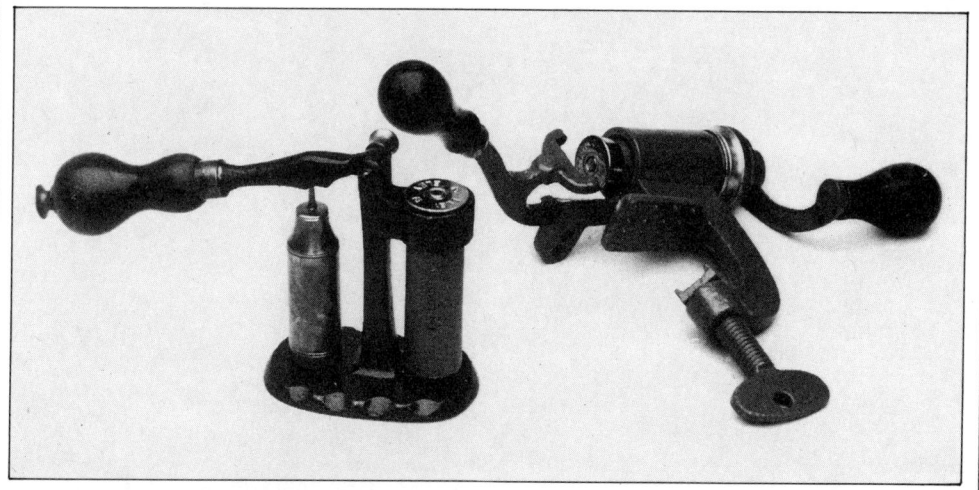

Bench closing machine (right) and de- and re-capper

of hours. Cartridges that are not corroded can be treated with an extra thin film of petroleum jelly on their brasses. Do not get the jelly on the paper.

Paper tubes on old cartridges may require refurbishing. They are often dirty through age and handling but with care some of the dirt can be removed. The secret is to use a piece of tissue paper whose texture is much softer than the cartridge tube. Moisten this with clean water and then gently rub off the dirt. Do not rub too long in one place. Some printing may stand a touch of washing-up liquid, but take care that the writing on the case is not erased. When all the dirt has been removed, let it dry for a few minutes and then rub it with a lump of beeswax and lightly polish with a very soft cloth.

The life goes out of the paper of many old cartridges through age, the paper becoming extremely soft. Never try to re-roll the turnover on such cases as the top of the cartridge may become screwed down and ruined. If you are determined to try to close the case the only answer is to try to put life back into the paper. To do this open the layers of paper at the top of the tube and work a non-staining gum into them. If the entire tube becomes saturated so much the better provided the outside layer is not affected. The case must now be carefully resized back to shape. Set it aside to dry when you are satisfied with the result. It will, however, be far from a total cure and even if you have strengthened the top the remainder of the case may be weak.

Practise techniques on cartridges you do not wish to keep so that you will not ruin any prized items. The three essentials in refurbishing are care, gentleness and patience.

Other Collectables

Though I have collected the odd cartridge box I have never become serious about them. The older boxes are often much sought after by collectors. They certainly make an eye-catching display and should not be dismissed as they can often be used as an exchange for a desired cartridge. It is also possible to glean information from them concerning old gun firms and their cartridges. Like old gunpowder tins, which are now difficult and expensive to acquire, old cartridge boxes are often attractively decorated. Very few cartridge collectors will part with the 100 size Eley Brothers boxes displaying the prize medals. These were produced in white and brown card. Kynoch & Co also sold their cartridges in 100 size wooden boxes with coloured printed paper on the outside depicting the lion's head trade mark and a cartridge. The boxes are very rare as they were ideal for planting seed and are often unrecognized as the paper has long since vanished. In some other countries, notably America, cartridge box collecting is taken very seriously and club meetings are held with prizes awarded for the best displays.

If you decide to start a cartridge collection and prefer only empty cases, do keep the overshot wads as these often carry the firm's name, though it may not appear on the side of the tube. These wads make the old manufacturers' display boards look attractive when they are inserted in the display. The days of collecting the display boards themselves are long since past as they now fetch three figures at auction.

Although it is not practical to collect the numerous headstampings on cartridges, several years ago I put together over 800 illustrations of overshot wads and headstamps of old British cartridges. If you are interested in reproducing headstamps, first make a foil rubbing by placing a small piece of foil over the headstamp and then gently rubbing from the primer cap out to the rim. This will create an impression sufficiently clear from which to draw. All I then do is to place a pair of compasses across them which brings them to double size. The headstamps illustrated on pages 30 and 31 will give a clear indication of their appearance.

Every serious cartridge collector should own a scrapbook which can be assembled from photocopied material taken from old albums and sporting books. Any information which will help to date cartridges is also valuable. Do not mutilate good books but have the advertisements copied, as they provide a great deal of useful information. Vintage catalogues are difficult to find and very expensive but always take the chance to copy them. Keep modern catalogues as one day these will be valuable sources of reference.

Pinfire Cartridges and Others

PINFIRE cartridges acquired their name from the pin that protrudes from the side of the base of the cartridge. The pinfire is, strictly speaking, a centre-fire cartridge as its percussion cap is centrally positioned in the base or head of the cartridge. The chief difference with the pinfire is that the cap is mounted on its side and is hidden from view. The pin in the side of the cartridges is equivalent to the striker pin in a hammer gun. Like the centre-fire, the pinfire originated in France. Many people believe the pinfire to be the forerunner of the centre-fire, but this is not so. Both types underwent their early development concurrently.

Many of the early pinfire cartridges loaded in Britain had cases imported from the Continent. When it soon became apparent that the breech-loader had taken precedence over the muzzle-loader three British firms started manufacturing pinfire cases as well as making centre-fire cases. These were Messrs Eley Brothers of 254 Gray's Inn Rd, London; George Kynoch Ltd of The Lion Works, Witton, Birmingham, and Frederick Joyce & Co Ltd of 57 Upper Thames St, London. The first pinfire cartridges of paper tubes were plain, but later a little printing was added. Not all pinfire cartridges were of paper; a few were marketed with full-length brass tubes.

There were several drawbacks to the pinfire cartridge. They were difficult to box up and for the user they posed a positive danger. They could not be carried in the pockets as a fall or blow might cause a disastrous explosion and the pins soon destroyed the linings of pockets. Care had to be taken when inserting the cartridge in the breech as the pin had to be firmly positioned in a slot in the end of the barrel prior to closing the gun. I was told that one of these cartridges dropped on its pin would explode like a Catherine-wheel in an unconfined space.

Once I was given a box of 50 I.C.I. white-cased 16-bore pinfire cartridges loaded with black powder and No. 6 shot. I set out to discover how dangerous these cartridges really were. Two were stripped down and dry sand replaced the powder. The wads and shot were replaced and the tops rolled in a closing machine. I dropped them and threw them at the ground but their caps never exploded. Finally, through wear and tear – not least to my pockets – they fell to pieces.

Personally, I doubt if the pinfire was as dangerous as we are led to believe; nevertheless I only experimented with one make – and that was of a later date than most. Like any other cartridges I treat them with respect.

Due to their various complications pinfires died a natural death, though they continued to find favour on the Continent rather longer than in Britain. I am not sure when the last British ones were made but they were still manufactured with the Eley-Kynoch I.C.I.-type headstamp as late as the 1930s in 12- and 16-bore white paper cases. Some of the last were loaded as blanks for alarm guns. These guns were clockwork operated and released

SOME VINTAGE CARTRIDGES

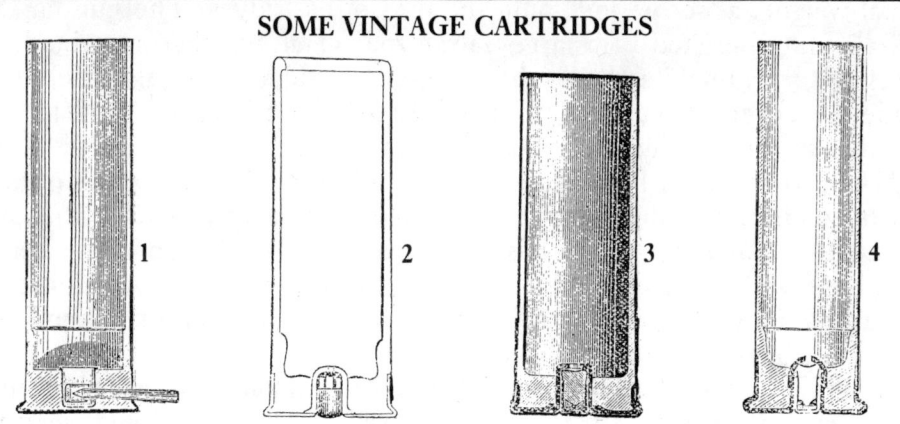

1. Pinfire case **2.** The Daw cartridge **3.** Joyce made, Bailey's patent **4.** An Eley Brothers centre-fire case **5.** A standard load in a paper-tubed case using black powder **6.** A paper-tubed case with a spreader or scatter load **7.** Full length brass ejector case with a crimp and compressed closure. Note smaller shot load

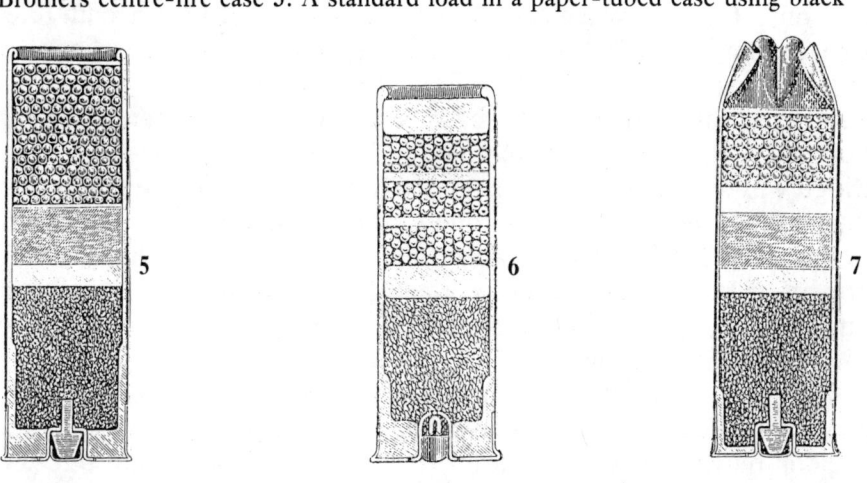

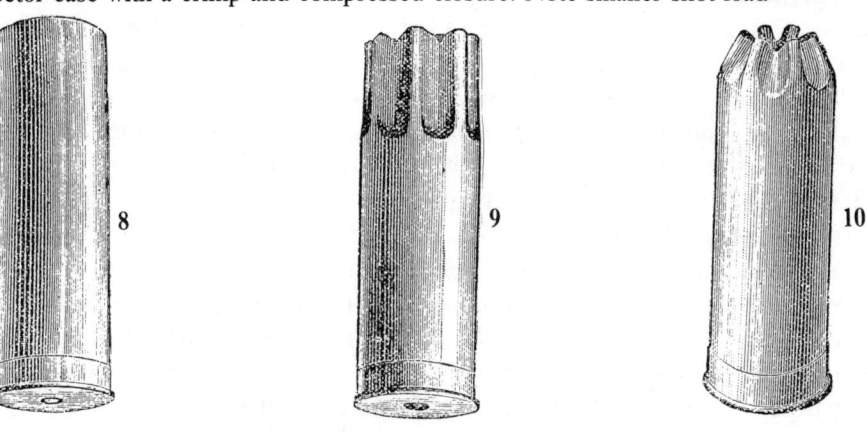

8, 9, 10. Kynoch's 'Perfect' full length brass case shown unclosed, closed with a corrugator and finally shown closed by crimping and compressing. An overshot wad is used to retain the load

metal weights at set periods onto the pins of cartridges. The gun had no barrels and consisted of a large galvanized metal box that contained the mechanism and held a row of these pinfire cartridges along its base. Once loaded, a cover protected it from the weather and it would be placed in position to act as a pigeon or poacher scarer.

When transporting pinfire cartridges in a box gently push pieces of hard plastic tubing over the pins. These can be cut a trifle longer than the pins. I use tubing cut from ends of insulation stripped from blue-covered telephone wire. When the wire is jointed the strips of plastic sheathing are discarded. However, any suitable plastic sheathing will do the job and give your cartridges extra protection in transit.

In 1852, Charles Lancaster, a well-known London Gunmaker, invented a centre-fire cartridge that could be used in a gun he had specially developed for it. This cartridge had a well-made rim for the extractor to operate on, a totally new concept. Lancaster's new cartridge had an internal base plate that had four holes grouped together in its centre. These were the flash holes. A thin layer of fulminate mixture was then sandwiched between this plate and the metal base of the head. The latter, when struck and dented, detonated the main charge. This cartridge looks like a pinfire cartridge without the pin – a sort of cross between a rimfire and a pinfire.

Once again credit for the invention of the true centre-fire cartridge must go to France. In 1853 a Mr Bellford took out a patent for a breech-loading centre-fire cartridge and Lancaster took this up but made no success of it. Then, in 1855, a Frenchman by the name of Pottet took out patents to modify this same cartridge.

In 1855 another centre-fire cartridge was being developed by François Eugene Schneider, who may have had family connections with George Schneider of Rue St Anne, Paris. It is known George took out several patents for improvements to breech-loading guns. F. E. Schneider's patent was then taken up in England by Mr George H. Daw of Threadneedle St, London, and thus was born the Daw cartridge. While this cartridge was being developed a long and complicated lawsuit took place between Daw and the Eley Brothers, which was won by the latter.

While the centre-fire and pinfire guns were being developed, the needle-fire gun had been evolved for a rifle in Germany. In Britain the needlefire became associated with a gunmaker, Mr Needham. He tried desperately hard to promote the system but it never caught on. The cartridge was constructed of paper and was fired by a thin spike, known as a needle, which penetrated the base and detonated the charge. The entire paper case was expelled from the barrel. Naturally needlefire cartridges are difficult to acquire! Although I have never seen a needlefire gun I believe that if a cartridge was not fired, or there was a misfire, it was extremely difficult to remove the cartridge from the chamber. A further reason for the system's unpopularity was that the fine needles used were prone to break and put the gun out of action. The cartridge's only real advantage was that, having no metal head, it was cheap to manufacture.

Mention must be made of electrical guns and their cartridges. Such a gun and its cartridge was shown at the Great Exhibition of Paris in 1867 by a M. le Baron. The gun was designed with a hollow stock which held a galvanic battery and an induction coil. The battery was housed in a gutta-percha waterproof cell. A connection was made with the battery when the trigger was pulled. The cartridge looked like a normal cartridge, but differed in its basic construction. In the interior of the base were two rings of copper let into the papier mâché base. Each was furnished with a point of wire and between these the electrical spark passed to detonate the charge. To meet the same challenge another electrical gun was made by Pieper of Liège in Belgium. In this case the shooter carried the battery himself and it was wired to a metallic shoulder-pad. Contact was made when the gun was raised to the shoulder-pad and met the metal buttplate. The idea behind this was to ensure the gun was safe whenever it was not being fired.

Little came of these strange inventions, yet they have their place in the story of the shotgun cartridge. Even today several cartridges are electrically detonated, usually to start engines.

Of all these old cartridges, the only types I have come across are some Eley Bros needlefire bullets and the Daw cartridges in 12- and 16-bore.

How Old is it?

THE QUESTION of age is one which has been put to me many times when I have been shown an old cartridge, but it is extremely difficult to provide accurate answers. Because it is impossible to give exact dates of manufacture for most cartridges I have chosen not to list dates for cartridges shown in this book.

The condition of a cartridge offers no clue to its age. Some cartridges look ancient merely because they have been badly stored. Others which are antique may look crisp and fresh because they have been stored in dark, dry conditions.

Knowledge of when the firm was in business is a starting point. This information can occasionally be obtained from local archives or the firm's records if still available. To do this one must have a good idea of the years when certain headstamps were in use. Many were employed for a number of years and one can only say that the cartridge was made between such and such dates. For example 'Special Smokeless' was a headstamp used in the 1920s, but it has also been seen on many foreign cases and is still used on Fiocchi cases today.

Occasionally the firm's address is printed on the cartridge and knowledge of when a gunmaker was in business at this particular address will help. Link this information to the headstamp and you may be able to pinpoint the date.

Additional information may be obtained from old catalogues or books and there are still people around whose memories may assist you.

In order to help I have included a page of headstampings with their approximate dates. These have been issued by the larger firms and I would emphasize that they are only approximate.

APPROXIMATE HEADSTAMP DATES OF THE BRITISH ISLES

E.B. No 12 LONDON	 ELEY No 12 LONDON	 KYNOCH No 12 BIRMINGHAM
1880–1895	1895–1919	1895–1919
 NOBEL No 12 GLASGOW	 ELEY N.I. No 12 LONDON	 ELEY 12 12 NOBEL
1912–1919	1920–1924	1924–1927
 KYNOCH 12 12 NOBEL	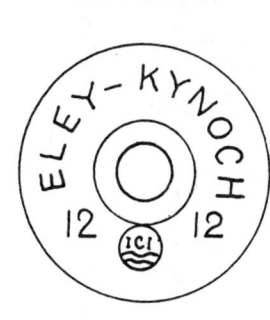 ELEY-KYNOCH 12 ICI 12	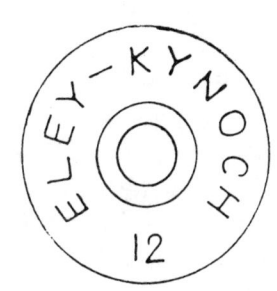 ELEY-KYNOCH 12
1924–1927	1926–1963	1963–1978

APPROXIMATE HEADSTAMP DATES OF AUSTRALIA

ELEY-KYNOCH I.C.I.A.N.Z. 12 12 MADE IN AUSTRALIA	 ELEY-KYNOCH 12 12 I.C.I.A.N.Z.	 ELEY-KYNOCH 12 ICI 12
1936–1949	1949–1956	1956–1965

The Cartridge List

THE FOLLOWING list records all the firms known to me in the British Isles that once made, loaded or sold their own brands of cartridges. As Northern Ireland and the Republic are closely linked I have included them both, though I have little information on Ireland.

The list covers the period from the first breech-loaders to the last paper tubes with rolled closures. After that the firms still in business changed their closure system to the modern star crimp. Although many of the brand names in the list continued with the new system, several firms changed some of their brand names, few of which will be found here.

Sadly the list of firms and their cartridges is far from complete as so much of the past has been lost to us. Information is always sought and, doubtless, the basics of a further compilation will gradually be gathered.

Where available, I have given information relating to each firm and occasionally I have supplied more than one example to illustrate the difference between an early and a later cartridge. Most brands were loaded in 12-bore, with a few in 12- and 20-bore, whilst often ·410 cartridges were given brand names such as the 'Midget'. It is not possible to list all the brands' sizes that were produced nor have I included details on ballistics.

The majority of measurements in the list are metric. Bl 8, for example, indicates that the brass length, including the cartridge rim, is 8 mm. Where the brass on a cartridge is a double-headed base, I have shown it so. For example: 8 + 17 totals 25, or 25 mm in length of brass head including the rim. The figures are to be read from the bottom to the top of a cartridge. The majority of old cartridge cases were about 65 mm long. This should be taken as standard unless otherwise stated. The ·410s were nearly always 50 mm and occasionally 65 mm.

It must be understood that whilst there were gunmakers and gunsmiths, there were also gunsmiths who called themselves gunmakers. Some actually made pistols and rifles or worked on them, whilst many dealt in fishing tackle. Nearly all the early firms loaded cartridges. I have included a heading 'Business' for each firm where this is known.

For most of the firms marketing a cartridge with its brass nearly, or right to the top, I have referred to it as an 'ejector'. In this list I have treated all these full-length cartridges as ejectors, the only exceptions being when they have been given brand names.

Whenever possible I have used descriptions of colours from catalogues. Where these are not available I have used my own eyes or information given to me by others. The first colour always refers to the overall colour of the paper tube and the second to the printing. Where there is no tube printing I have used the word 'nil'. For example: Black/silver or purple/nil. I have used the same system to describe colours on overshot or top wads. This is then followed by the wording and the shot size.

Cartridge Code

B   Used under heading Pr to indicate a brass material used for the percussion or primer cap.

Bl   The brass length or length of the cartridge head, including the rim.

C   Under heading Pr to indicate a copper material used for percussion cap.

De   Decor. A picture of game, a crest or trade mark etc.

DECOR

← On, NOBEL - GLASGOW case.
Example. E. CHAMBERLAIN of ANDOVER & BASINGSTOKE.
Note — No right rear foot.

On, KYNOCH. BIRMINGHAM case →
Example. A. BATES of CANTERBURY.
ARTHUR CONYERS of BLANDFORD.

← On, ELEY-KYNOCH. I.C.I. case.
Example. H. & R. SNEEZUM of IPSWICH.
Note — Seen on, SPECIAL SMOKELESS.

On, ELEY-KYNOCH I.C.I. case. →
Example. JOHN ANDERSON & SON. of MALTON.

← On, ELEY-KYNOCH. I.C.I. case.
Example. R. & E. POTTER of THAME.

On, ELEY-KYNOCH. I.C.I. case →
Example. HARRY HIGGINS. of TENBURY WELLS.

Subject to variations

DECOR

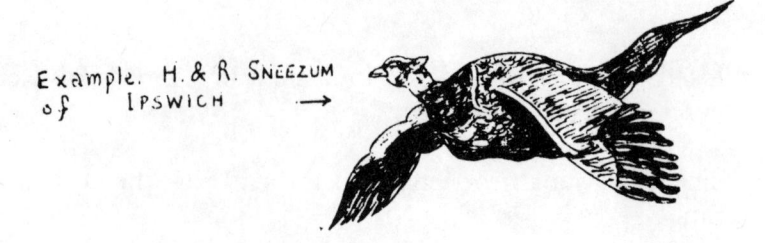

Example. H. & R. SNEEZUM of IPSWICH →

Example. WHITEMAN BROS of WORCESTER ←

Example. HOWARD. A. DAVIES of WINCHESTER →

Example. LANGLEY & LEWIS of LUTON. ←

Example. TOM LAW of CASTLE DOUGLAS →

Example. R. ROBINSON of HULL. ←

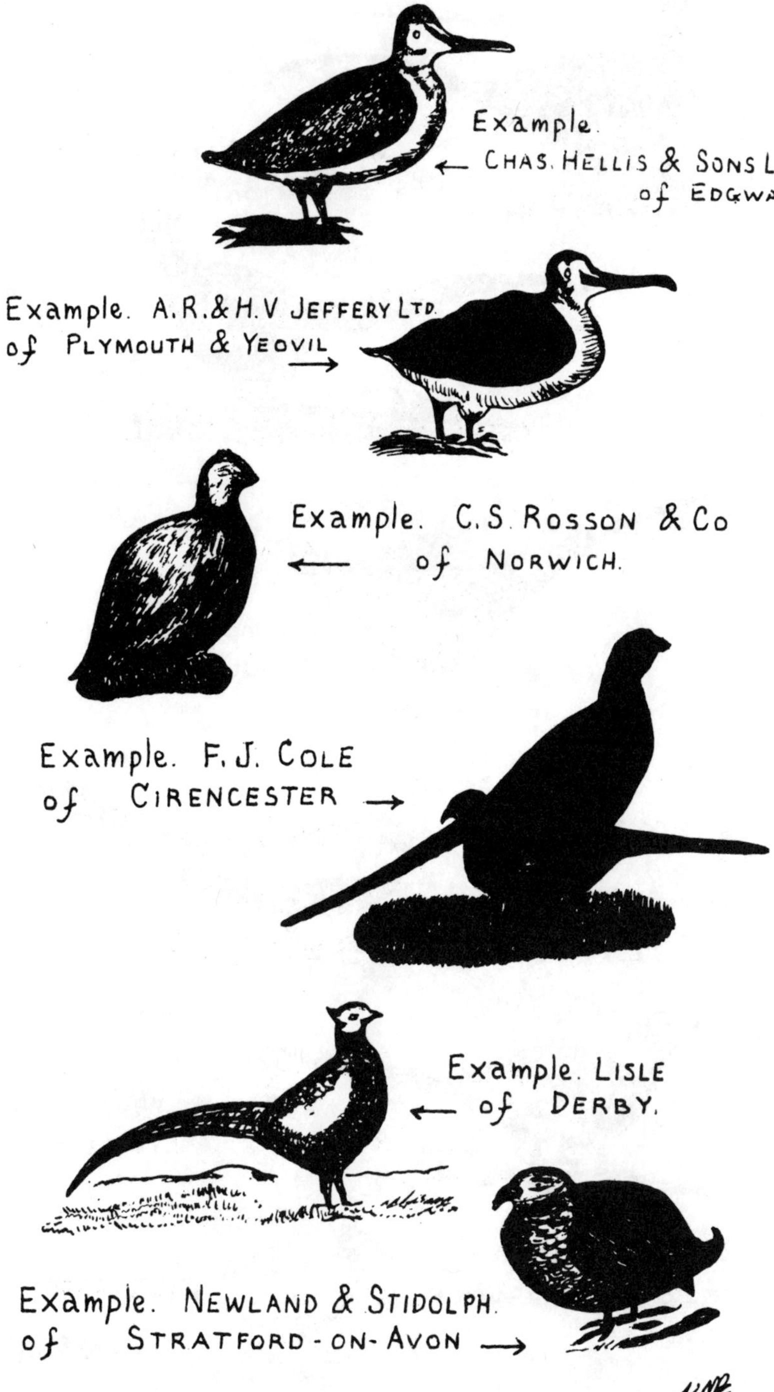

DECOR

Example. CHAS. HELLIS & SONS LTD. of EDGWARE

Example. A.R. & H.V JEFFERY LTD. of PLYMOUTH & YEOVIL

Example. C.S. ROSSON & Co of NORWICH.

Example. F.J. COLE of CIRENCESTER

Example. LISLE of DERBY.

Example. NEWLAND & STIDOLPH. of STRATFORD-ON-AVON

DECOR

← Two variations of The "CHAMPION." →

← Example. JOHN H. GILL & SONS. (LEEMING BAR).

Example. J.J. HOPKINS. of LEIGHTON BUZZARD. →

← Example. W. JEFFERY & SON of PLYMOUTH.

Example DARLOW & Co of NORWICH →

F	Front view when describing game, crest etc.
Fl	Viewed from front left quarter when describing game, crest etc.
Fr	Viewed from front right quarter when describing game, crest etc.
Ga	Gauge or bore size.
Kn	Knowledge. Information about the firm or cartridge.
L	Subject in question when viewed from left. Also used under heading Pr to indicate a large cap diameter.
M	Under heading Pr to indicate Medium cap diameter.
N	Under heading Pr to indicate no insert cup is used. Cap pressed directly into head.
Or	Origin. Country in which case was made.
Pr	Primer cap. Details where known.
R	Subject in question when viewed from the right.
S	Under heading Pr to indicate a small-diameter cap.
St	Stamping. Marking of the headstamp. (See illustrations on pages 30 and 31.) (Reversed). Headstamp pressed out from underneath, markings therefore raised.
Tc	Colour of the paper tube followed by printing colour.
Wd	Overshot wad or top wad. Colour, wording and shot size.

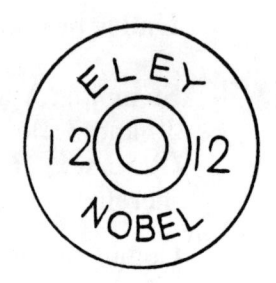

ABBREVIATIONS OF THE MORE COMMON HEADSTAMPINGS USED IN THE LIST

ABBREVIATIONS OF THE MORE COMMON HEADSTAMPINGS USED IN THE LIST

Firms of the British Isles and their Cartridges (in alphabetical order)

ACCLES ARMS, AMMUNITION & MANUFACTURING CO: Perry Bar, Birmingham (Midlands).
Business: Gun and cartridge makers.
Cartridges:
Example: Unnamed. Ga 12; Tc blue/nil; Pr MCI; St Accles Limited № 12 Birmingham; Or British.

This example was taken from a box of six ready primed cases that exist in a private cartridge collection. The other cases were identical except for their tube colours. They were cream, green, orange, pink and red.

This firm was founded at the time that Grenfell & Accles Limited was dissolved. It only operated between 1896 and 1899.

ADAMS: Crediton, Devon.
Business: Ironmonger.
Cartridges:
Example: Unnamed. Ga 12; Tc off-white/black; De Running rabbit; Bl 8; St MGB; Or British.

ADAMS: Littleport, Cambs.
Cartridges: No details recorded.

ADAMS & CO: Finsbury, London.
Cartridges: No details recorded.

ADGEY: Belfast, Northern Ireland.
Cartridges: De Luxe; The Favourite; Special de Luxe; The Universal.
Example: The Favourite. Ga 12; Tc red/black; Kn Loaded in Ireland with a high-grade smokeless powder; St Made In England; Or British.

H. ADKIN & SONS: Bedford, Beds.
Business: Gunmakers.
Cartridges: 20 Gauge; The Ajax; The Demon; The Reliance; Special Loading.
Example: 20 Gauge. Ga 20; Tc buff/black or dark blue; De Coat of arms with a castle and a background of an outstretched eagle; Bl 7; Pr MCI; St ICI; Or British.

T. ADSETT & SON: 101 High St. Later at 90 High St, Guildford, Surrey.
Business: Gunmakers.
Cartridges:
Example one: Unnamed. Ga 12; Tc burgundy/black; Kn 101 High Street. Smokeless cartridge; Bl 8; Pr LCI; St MGB; Wd white card/pencilled on, Buck Shot; Or British.
Example two: Unnamed. Ga 12; Tc Gastight brick red/black; Kn 90 High Street. Water resisting. Steel lined; Bl 16; Pr MCI; St ICI; Wd white card/nil; Or British.

AGNEW & SON: 79 South St, Exeter, Devon.
Business: Gunmakers.
Cartridges: The Devonia.
Example: Unnamed. Ga 12; Tc orange/black; Bl 8; Pr MCI; St KB; Or British.

H. ESAU AKRILL: 18 Market Place, Beverley, Yorks (Humberside).
Business: Gunmaker:
Cartridges: The Collector; The County; The Holderness; The Special; The Universal Cartridge.
Example: The Collector. Ga ·410; Tc Indian red/black; Kn Produced in 2 inch and 2½ inch case length. Loaded by Akrill, Beverley; Bl 7; Pr MCI; St Eley-Kynoch ·410; Or British.

E. ALDRIDGE: Hyde Park Corner, Ipswich, Suffolk.
Cartridges: The Crown; The Hyde Park; The Anglian.
Example one: The Crown. Ga 12; Tc purple/gold; De Large crown; Bl 8; Pr MCI; St SS; Wd white card/black. Special · Smokeless · 5; Or Foreign.
Example two: The Anglian. Ga 12; Tc purple/silver; Kn High velocity. Specially loaded by E. Aldridge; Bl 16; Pr MCI; St Smokeless 12 12 Gastight. (Inner) RWS Nuremberg; Wd Colour not recorded. Aldridge · Ipswich · 4; Or German.

ALEXANDERS: Fordingbridge, Hants.
Cartridges: The Fordingbridge.
Example: The Fordingbridge. Ga 16; Tc lichen green/black; De A picture of the town bridge showing its six arches; Bl 8; Pr MCI; St EL; Or British.
This item was an unused ready capped case. A similar case in the same colour and with the same printing has been seen in Ga 12.

ALEXANDER & DUNCAN: Leominster & Hereford.
Cartridges:
Example: Unnamed. Ga 12; Tc red/black; De Cock pheasant standing in grass with a raised tail, R; Bl 8; Pr MCI; St EL; Or British.

ARTHUR ALLAN: 3 West Nile St, Glasgow, Lanarks.
Business: Gunmaker.
Cartridges: Super A.A.; Three Star.
Example: Three Star. Ga 12; Tc red/black; Bl 8; Pr MCI; St ICI; Wd yellow/black. Special · Smokeless · 5; Or British.

ALLTHAM & SON: Penrith, Cumberland (Cumbria).
Cartridges: Eley's Pegamoid:
Example: Eley's Pegamoid. Ga, 12; Tc, brown/black; De The EBL Shield Trade Mark with outlined lettering and rope-type edging; Kn All the tube printing is that of Eley. Steel lined; Bl 16; Pr MCI; St Alltham & Son. Penrith . Nº 12 Eley; Or British.

ALPASS & BAKER: Wiveliscombe, Somerset.
Business: Ironmongers.
Cartridges: I have been told of their top wad. This was White card/red. It was loaded into an Eley Brothers plain case.

JOHN ANDERSON & SON: Malton, Yorks.
Business: Gunmakers and sports dealers.
Cartridges: The Derwent; The Eclipse; The Rabbit:
Example: The Rabbit. Ga 12; Tc poppy red/black; Bl 8; Pr LCI; St Jas. R. Watson & Cº 12 12 London; Or Belgium.

ARMSTRONGS: Carlton Hill, Nottingham, Notts.
Business: Gunmakers.
Cartridges: The Sherwood.

ARMSTRONGS (SPORTING GUN DEPOT): Also, ARMSTRONG & CO: Newcastle-Upon-Tyne, Nthmb. (Tyne & Wear).
Cartridges: THE A.C.C.; Gastight; Pressure Reducing Case.
Example: Gastight. Ga 12; Tc brownish orange/black; Kn The words, Sporting Gun Depot, in old English; Bl 16; Pr MCI; St Armstrong & Cº 12 12 Newcastle-Upon-Tyne; Or Not known.

ARMY & NAVY CO-OPERATIVE SOCIETY: Later changed to, ARMY & NAVY STORES: LONDON.

The crests on The Army & Navy C.S. Ltd and the later Army & Navy Stores Ltd

Business: London stores and gunmakers.
Cartridges: The Coronation Cartridge; The Every Day; The Every Day Nitro; The Fureka; Gastight; The Nitro; Pegamoid: The Reliable; The Victoria.
Example one: The Coronation Cartridge. Ga 12; Tc russet brown/ black; De and Kn Army & Navy C.S. Ltd. The wording, 'Coronation Cartridge' around the top of the tube with a crown between them. Below this were crossed flags; Bl 25; Pr MCI; St EL; Or British.
Example two: The Nitro. Ga 12; Tc middle blue/ black or dark blue; De Large round crest (See page 33); Kn, Army & Navy Stores Ltd, London; Bl 8; Pr MCI; St ICI; Wd white card/black. Lethal Ball; Or British.
The Victoria Cartridge

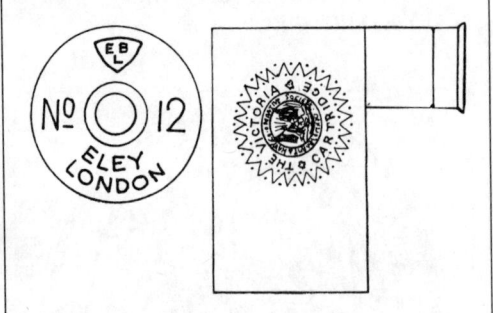

The above drawing was taken from an illustration in an early Army & Navy Co-operative Society's catalogue.
Advertised in 12-bore, the wording read: 'A very best paper case, half brass covered, made especially for the Society, and loaded with great care by Messrs. Eley Bros. with Bulk Nitro Powders and the very best materials. Per 100 . . . 9/6. Recommended in 12-Bore only.'
Case and wad colours are not known.

HENRY ATKIN: 41 Jermyn St. Later at 88 Jermyn St, London, S.W.1.
Business: Gunmaker.
Cartridges: The Covert; Ejector; The Ever Ready; The Gem; Pegamoid; The Raleigh:
Example one: Pegamoid. Ga 12; Tc purple/black; Kn Specially loaded by Henry Atkin, Ltd. 41 Jermyn Street. Made in Great Britain from Pegamoid Brand paper; Bl 16; Pr MCI; St Atkin · 41 Jermyn St · N⁰ 12 Eley; Wd red/black. H.Atkin; Special Loading. 5½; Or British.
Example two: The Gem. Ga 12 Tc grey/black; Kn Cartridge name in long-hand. 88 Jermyn Street. Telephone Whitehall 4644. Telegrams

'Atkinelio, Phone, London'; Bl 8; Pr MCI; St ICI; Or British.

ATKINSONS: 31 Oxford St, Swansea, Glam.
Business: Gun dealers.
Cartridges: The Grand Finale.

T. ATKINSON & SONS: 19a Stricklandgate, Kendal, Westm (Cumbria). Also at, Lancaster, Lancs.
Cartridges: The Ajax; The Kendal Castle; The Kent; The Kendal; The Reliable.
Example one: Atkinson's Ajax. Ga 12; Tc purple/ yellow; Kn Lancaster & Kendal; Bl 8; Pr MCI; St Atkinson N⁰ 12 Lancaster & Kendal. Eley; Wd salmon pink/black. Atkinson · Lancaster · (Inner) Schultze. 5; Or British.
Example two: The Kent. Ga 12; Tc Eau-de-nil/ dark blue; Kn T. Atkinson & Sons. 19a Stricklandgate, Kendal. Phone 300; Bl 8; Pr MCI; St ICI; Wd white card/red. Smokeless. 5; Or British.

K. ATTRILL: Pyle St, Newport, Isle of Wight.
Cartridges: It is known that he loaded and sold his own cartridges.

T. C. AUSTIN: Ashford, Kent.
Business: Gunmaker.
Cartridges: Eley Case.
Example: Eley Case. Ga 12; Tc dark red/black; De The EBL Shield Trade Mark with black lettering and rope-type edging; Kn Specially loaded by T. C. Austin. Gunmaker, Ashford; Bl 8; Pr MCI; St EL; Wd white card/black. Shot size only, 5; Or British.

AVERILL & SON: Evesham, Worcs.
Cartridges: Averill's Express.
Example: Averill's Express. Ga 12; Tc ruby red/ black; Kn Smokeless cartridge; Bl 8; Pr MCI; St Averill & Son N⁰ 12 Evesham; Wd salmon pink/black. Averill & Son · Evesham · (Inner) Smokeless 5; Or Not known, believed to be British.

34

BACON & CURTISS: Poole, Dorset.
Cartridges:
Example: Unnamed. Ga 12; Tc off-white/nil; Bl 8; Pr MCI; St EN; Wd Bacon & Curtiss · Poole; Or British.

BAGNALL & KIRKWOOD: 31 Westgate Rd, Newcastle-upon-Tyne, Nthmb (Tyne & Wear).
Business: Gunmakers and fishing tackle experts.
Cartridges: The Pointer; The Setter.
Example: The Setter. Ga 12; Tc pink/black; De Setter dog, L; Kn Late with W. R. Pape; Bl 8; Pr MCI; St ICI; Wd white card/red. Smokeless. 5; Or British.

E. BAILDHAM & SON: Stratford-upon-Avon, Warwickshire.
Business: Ironmongers.
Cartridges: The Duck Fowler; The Extra; The Standard.

CHARLES S. BAILEY: Waltham Abbey, Essex.
Business: Inventor and possibly a cartridge manufacturer.
Cartridges: In the year 1882 he patented a gas check for central-fire cartridges. An extra-thin layer of metal hid the primer cap. This patent was later used by Frederick Joyce & Co.

W. R. BAILEY: Congresbury, Somerset.
Business: Agricultural and ironmongers.
Cartridges:
Example: Unnamed. Ga 12; Tc yellow/black; Kn This was a Mullerite cartridge with extra tube printing including W. R. Bailey's name; Bl 8. No other details recorded.

C. T. BAKER: Holt, Norfolk.
Business: Ironmonger.
Cartridges: They sold their own brand cartridges in competition with Joseph Baker & Son. So far no cartridge is known to have survived though many were handed in to the police and destroyed.

F. T. BAKER: 29 Glasshouse St, London. W.
Business: Gun and rifle maker.
Cartridges: Baker's Best; Ejector.
Example: Baker's Best. Ga 12; Tc burgundy/silver; Bl 16; Pr MCI; St F. T. Baker. London. № 12 Eley; Or British.

J. T. BAKER: 103 Victoria Rd, Darlington, Co Durham.
Business: Gun and fishing tackle dealer.
Cartridges: Sipe Smokeless.
Example: Sipe Smokeless. Ga 16; Tc dark red/black; Kn This cartridge was a Sipe Smokeless by Le Personne & Co; Bl 7; St Lepco. Made in Italy; Wd Baker. Darlington; Or Italian.

JOSEPH BAKER & SON: Fakenham, Norfolk.
Business: Ironmongers and gunsmiths.
Cartridges: Baker's Special.
Example: Baker's Special. Ga 12; Tc pale greyish green/black; Kn Special Smokeless cartridge. Telephone, Fakenham 22. Telegrams, Baker, Fakenham 22; Bl 8; Pr MCI; St ICI; Or British.

W. E. BAKER: Tavistock, Devon.
Cartridges: Kynoch Witton Brand Case.
Example: Kynoch Witton Brand Case. Ga 12; Tc brown quality/nil; Bl 8; Pr SCI; St W. E. Baker № 12 Tavistock; Or British.
The example is prior to 1899. It is one from a sample box sent by G. Kynoch & Co to the U.S.A.

BALLS: Bungay, Suffolk.
Cartridges: I have been told that many years ago, cartridges were loaded on the premises.

J. C. BANFIELD & SONS: Tenbury Wells, Worcs.
Business: Ironmongers.
Cartridges: Unitro Cartridge Case.
Example: Unitro Cartridge Case. Ga 12; Tc orange/black; De Cock pheasant standing with tail outstretched, L; Kn Ironmongers; Bl 8; Pr MCI; St ICI; Or British.

BAPTY & CO: London.
Business: Arms retailers.
Cartridges:
Example: Unnamed. Ga 12; Tc black/nil; Bl 13, with diagonal grooves running round the brass head; Pr MBI; St within a recess, № 12 U S Climax; Wd white card/pink. Bapty & C°. London. 5; Or U.S.A. Note: Name on wad only.

H. W. BARFORD & CO: 14, 15, 16, Bishop St, Coventry, Warwicks (W. Midlands).
Cartridges: Special Imperial.
Example: Special Imperial. Ga 12; Tc gastight brick red/black; De Cock pheasant standing in grass with lowered tail, L; Kn smokeless cartridge. Steel lined; Bl 16; Pr MCI; St ICI; Wd yellow/black. Smokeless · Diamond · 5; Or British.

C. H. BARHAM: 95 Tilehouse St, Hitchin, Herts.

A crest from C. H. Barham

Business: Gunmaker.
Cartridges: The Challenge; The Comet Cartridge; The Hert's Cartridge.
Example: The Challenge. Tc red/black; De Cock pheasant walking in grass with head and tail high, L; Kn loaded by C. H. Barham, Gunmaker Hitchin. Phone, Hitchin 498; Bl 8; Pr MCI; St ICI; Wd white/card/red. Barham * Hitchin * ; No shot size; Or British.

BARKERS: Corner House, Huddersfield, Yorks.
Cartridges: The De Luxe.
Example: The De Luxe. Ga 12; Tc black/silver; De Hen pheasant walking with tail high, L; Kn Special smokeless. British loaded; Bl 16; Pr MCI; St Smokeless 12 12 Gastight; Wd white card/black. Special · Smokeless · 5; Or Foreign.

A. BARNES: Ulverston, Lancs (Cumbria).
Business: Gunmaker.
Cartridges: The Referendum.
Example: The Referendum. Ga 12; Tc purple/yellow; De Cock pheasant standing in grass with tail horizontal, L; Bl 16; St A. Barnes № 12 Ulverston; Or Not known.

G. J. BARNES & SON: 29 Church St, Calne, Wilts.
Business: Ironmongers.
Cartridges:
Example: Unnamed. Ga 12; Tc tan brown/black; Kn Kynoch's Perfectly Gas-tight Cartridge; Bl (Double head) 7 + 16; Pr SCI; St Barnes № 12 Calne; Wd white card/black. (Cluster type) 6; Or British.

JOHN BARNES: Ayr, Ayrshire.
Cartridges: The Challenger.

BARTRAM: Braintree, Essex.
Cartridges: Bartram's Hard Hitters.
Example: Bartram's Hard Hitters. Ga 12; Tc dull orange/black; Kn Smokeless powder; Bl 8; Pr MCI; St EL; Wd white card/black. Bartram. Braintree. 6; Or British.

G. J. BASSETT: 4 Swan St, Petersfield, Hants.
Business: Ironmonger.
Cartridges: The Champion.
Example: The Champion. Ga 12; Tc mauve/silver; De Rabbit outstretched over a grass surface, L; Kn Special smokeless. British hand loaded. Phone Petersfield 238; Bl 7; Pr LCI; St SS; Or Foreign.

GEORGE BATE: 132 Steelhouse Lane, Birmingham (Midlands).
Business: Gunmaker.
Cartridges: The Game; The Imperial; The Leader.
Example: The Game. Ga 12; Tc middle blue/black or dark blue; De Cock pheasant standing in grass with tail in a slight downward position, L; Kn Gastight. Water resisting; Bl 16; Pr MCI; St ICI; Wd bluish grey/black. Empire. 6; Or British.

A. BATES: 22 Sun St, Canterbury, Kent. Also depots at Sturry & Whitstable, Kent.
Business: Gunmaker.
Cartridges: The Challenge; The Rabbit.
Example: The Challenge. Ga 12; Tc dark red/black; De Rabbit running over a young foxglove plant, FL; Bl 8; Pr MCI; St KB; Wd pale orange/black. A. Bates+Canterbury+ (Inner) Schultze 6; Or British.

E. R. BATES & SONS: 3 George Gate. Also at 71 Burgate St, Canterbury, Kent.
Cartridges: The Challenge.

GEORGE BATES: Eastbourne, Sussex.
Business: Gunmaker.
Cartridges: The Eastbourne; The Reliable.
Example: The Eastbourne. Ga 12; Tc brownish orange/black; De Coat of arms; Kn Schultze powder; Bl 10: No other details recorded.

BAYS & CO: Wood St, Swindon, Wilts.
Business: Stores.
Cartridges:
Example: Unnamed. Ga 12; Tc brownish orange/black; De Cock pheasant standing in grass with tail outstretched, L; Kn Bays & Co. Swindon; Bl 7; Pr MCI; St NG; Or British.

H. BECKWITH: London.
Cartridges: A cartridge collector told me that a 15-bore pinfire had been seen.

BEESLEY: London.
Cartridges: Kynoch Grouse Ejector.
Example: Unnamed. Ga 12; Tc crimson/black; De Crest of The Prince of Wales feathers; Kn By appointment; Bl 16; Pr MCI; St ICI; Wd yellow/black. Beesley Loading. 4; Or British.

BELCHER: Wantage, Berks (Oxon).
Business: Ironmonger.
Cartridges: Information given was that at one time cartridges were loaded on the premises.

BELLOW & SON: Leominster, Hereford, Tenbury Wells & Bromyard.
Cartridges: Special Cartridge.
Example: Special Cartridge. Ga 12; Tc yellow/black; Kn This cartridge had Mullerite printing with extra tube printing for Bellow & Son; Bl 8. No other details recorded.

A. BENHIGNO: 12–14 High St, Peebles, Peebleshire.
Business: Retailer in guns and cartridges.
Cartridges:
Example: Unnamed. Ga 12; Tc mauve/silver; Bl 8; Pr MCI; St SS; Or Foreign.

G. W. BENNETT: Blackpool, Lancs.
Cartridges: Mullerite Yellow Seal.
Example: Mullerite Yellow Seal. Ga 12; Tc yellow/black; Kn This cartridge had Mullerite printing with extra tube printing for G. W. Bennett, Blackpool; Bl 8; Pr LCI; St SS; Or Not known.

J. BENTLEY: 309 Halifax Rd, Liversedge, Yorks.
Business: Cartridge loading and sales.
Cartridges: The Croft.
Example: The Croft. Ga 12; Tc orange or crimson/black; De A small round-breasted cock pheasant. No ground with tail down while standing, L; Bl 8; Pr MCI; St MGB; Wd light blue/black. Special · Smokeless · 4; Or British.

Joe Bentley marketed cartridges that used Greenbat Powder and Greenwood & Batley cases.

BEVAN: address unknown.
Cartridges: The Dreadnought.
Example: The Dreadnought. Ga 12; Tc stone white/black; De Cock pheasant in grass with raised tail, L; Kn Smokeless cartridge; Bl 8; Pr MCI; St NG; Wd white card/black. Smokeless 5; Or British.

BEVAN & EVANS: Abergavenny, Mon (Gwent).
Cartridges: The Abergavenny Ace.
See, Bevan & Pritchard, below.

BEVAN & PRITCHARD: Abergavenny, Mon (Gwent).
Cartridges: The Abergavenny Ace.
Example: The Abergavenny Ace. Ga 12; Tc cream/not recorded; De A shield with two crescents forming the top, within is a cutlass type centre with two outer forming as three feathers. Below is a scroll with the wording, 'The Abergavenny' within it. This is followed by the words, Ace and Smokeless; Kn Sold by Bevan & Pritchard; Bl 8; Pr MCI; St KB; Wd Colour not recorded. Evans · Abergavenny · (Inner) Smokeless 7; Or British.

BISSET: London.
Cartridges:
Example: Unnamed. Pinfire. Tc dark green/nil; Bl 6; St (Reversed). St Bissett "12" London; Wd Plain cork or paper; Or Not known.

J. BLACK: Bollington, Ches.
Business: Gun dealer.
Cartridges: The Bollin.

C. G. BLACKADDER: Castle Douglas, Kirkcudbright.
Cartridges: The Black Douglas.
Example: The Black Douglas. Ga 12; Tc pink/black; Kn Loaded by C. G. Blackadder; Bl 8; Pr MCI; St ICI; Wd white card/black. C. G. Blackadder · Castle Douglas · 4; Or British.

J. BLAIN: Carlisle, Cumberland (Cumbria).
Business: Ironmonger.
Cartridges: No cartridge details available.

BLAKE BROTHERS: Ross-on-Wye, Herefordshire.
Business: Ironmongers.
Cartridges: The Wye Valley.

JAMES BLAKE: 12 Square, Kelso, Roxburgh.
Cartridges: The Roxburgh.
Example: The Roxburgh. Ga 12; Tc pink/dark blue; De Cock pheasant standing in scrub with tail horizontal, L; Kn Smokeless cartridge; Bl 8; Pr MCI; St ICI; Or British.

J. BLANCH & SON: 29 Gracechurch St. Also at 4 Bishopgate Churchyard, London.
Business: Gunmakers.
Cartridges: Ejector.
Example: Ejector. Ga 12; Two-piece brass case with a maroon inner paper tube; Bl 7 + 53; Pr MCI; St J. Blanch & Son No 12 London. (Inner) Kynoch's Patent Grouse No.2090; Wd, pale yellow/black with the print forming a background for yellow wording. Blanch · Schultze · 6; Or British.

This business was later taken over by Alfred Davis.

THOMAS BLAND & SON: 430 West Strand. Later at 2 King William St, London.
Business: Gunmakers.
Cartridges: The address changed in April, 1900. All the cartridges that I have seen had no brand names.

R. BLANTON: Market Place, Ringwood, Hants.
Business: Gunmaker.
Cartridges: The Competitor; The Imperial.
Example: The Competitor. Ga 12; Tc, light purple/silver; De Hen pheasant walking on pine needles with raised tail, L; Kn British loaded. Phone, Ringwood 223; Bl 8; Pr MCI; St SS; Wd yellow/black. Special · Smokeless · 5; Or Foreign.

BLYTHE & WRIGHT: Station Rd, Sheringham, Norfolk.
Business: Ironmongers.
Cartridges: This firm loaded cartridges many years ago.

G. EDWARD BOND & SON: 1 Market Place, Thetford, Norfolk.
Business: Gunmakers.
Cartridges: Eley's Ejector; The Invincible; Pegamoid.
Example: Pegamoid. Ga 12; Tc brown/black; Kn Eley Gas-tight Waterproof Cartridge Case. From Pegamoid Brand paper. Specially made for Bond & Son. Gunmakers, Thetford. Steel liner; Bl 16; Pr MCI; St Bond & Son No 12 Thetford. Eley; Or British.

J. S. BOREHAM: Colchester, Essex.
Cartridges: Kynoch Witton Brand Case.
Example: Kynoch Witton Brand Case. Ga 12; Tc brown quality/nil; Bl 8; Pr SCI; St Boreham No 12 Colchester; Or British.

BOSS & CO: 13 Dover St. Later at 41 Albemarle St. Then 13–14 Cork St, London.
Business: Gunmakers.
Cartridges: Ejector; High Velocity; Special. Some other brands were catalogued as Brown Brand; Green Brand; Orange Brand: This by the colour of their tubes.
Example: Special. Ga 12; Tc red/black; Kn Ga 12 only. Sold in 1936 at 18/- per 100; Bl 16; Pr

LCI; St ICI; Wd red/black. Boss & Co · London · 5½; Or British.

Boss & Co's catalogue for 1935–1936 also showed a cartridge coloured a bluish grey. The later coloured Boss & Co's cartridges were coloured pink/black.

Boss & Co. Special. Case colour: Signal red with black print. Wad colour as illustrated: French blue with black print.

Note the extra-large copper primer cap.

CHARLES BOSWELL: 126 The Strand. Later at 15 Mill St Hanover Square, London.
Business: Gun and rifle maker.
Cartridges: Special Express.
Example: Unnamed. Ga 12; Tc maroon/silver; Kn Diagonally printed, Kynoch's Patent Perfectly Gastight Cartridge Case; Bl (Double head) 6 + 17; Pr -CI; St C. Boswell № 12 126 Strand; Wd white card/red with the print forming a background for white wording. Boswell · 126 Strand · 6; Or British.

JOHN BOWEN: Carmarthen (Dyfed).
Cartridges: Myrddin.
Example: Myrddin. Ga 12; Tc pale yellow/black; De Cock pheasant with head held high L; Kn Loaded Schultze; Bl 7; Pr MCI; St Schultze № 12 London. CHMS; Or British.

BOWERBANKS: Penrith, Cumberland (Cumbria). Also at Kirkby Stephen, Westm (Cumbria).
Cartridges:
Example: Unnamed. Ga 12; Tc pink/black or dark blue; De Cock pheasant standing; Bl 8; Pr MCI; St ICI; Or British.

BRADDELL & SON: Arthur Square, Belfast, Northern Ireland.
Business: Gunmakers.
Cartridges: The Castle; De Luxe Special; Ejector; The Empire; The J.B.; The Mors; The Special; Special Gastight; The Victory.
Example: The Special. Ga 12; Tc primrose yellow/(Two-colour print) blue-black and red; De Trade Mark. This is the only place where the red is used. It is a gloved red human left hand pointing skywards with the monogram JB in its palm. The rest of the printing is blue-black; Kn Water resisting. Gastight. Metal lined. Made in Great Britain; Bl 16; Pr MCI; St ICI; Wd light yellow/black. Braddell's Special 8; Or British.

Braddell's also catalogued and sold the following cartridges; F.N. All-metal H.V.; Nobel's Sporting Ballistite; Noneka; Alphamax 2½ and 3 inch; Bonax Ga's 12 and 16; Grand Prix Ga's, 12 and 16; Maximum; Primax Ga's 12 and 16; Eley Smokeless; Special Trap-shooting; Eley Velocity; Westminster Ga's 12 and 16; Zenith; Wildfowling; 20th Century; 20-Bore; 28-Bore Gastight; 10-Bore; Also pinfires: This is typical of many of the large firms that also sold other brands of cartridges as well as their own.

BRIND GILLINGHAM: Ock St, Abingdon, Berks (Oxon).
Business: Ironmongers.
Cartridges: Information given to me that cartridges were once loaded.

R. BROADHURST: Smithford St, Coventry, Warwickshire (Midlands).
Business: General ironmonger and furnishing.
Cartridges: Eley's Special Smokeless.
Example: Eley's Special Smokeless Sporting Powder. Ga 12; Tc pink/black with an additional purple oval stamping; De and Kn The stamping with the wording 'R. Broadhurst. General furnishing. Ironmonger. Smithford Street, Coventry.'; Bl 8; Pr -CI; St EB. Nitro № 12 Eley; Or British.

BROCK'S EXPLOSIVES: Hemel Hempstead, Herts.
Business: Pyrotechnic manufacturers.
Cartridges: Bird Scaring Cartridge.
Example: Bird Scaring Cartridge. Ga 12; Tc orange/black; 2 inch (50 mm); Kn Ejecting a humming star; Bl 8; Pr MBI; St Eley-Kynoch 12; Or British.

JOHN BROMLEY & CO: Wellington, Newport, Shifnal & Bridgnorth, Salop.
Business: Ironmongers.
Cartridges:
Example: Unnamed. Ga 12; Tc middle blue/black; De Rabbit running over a plant, FL; Kn John Bromley & Co (Wellington) Ltd; Bl 8; Pr MCI; St ICI; Wd yellow/black. Smokeless · Diamond · 6; Or British.

E. J. BROWN & CO: Rotherham, Yorks.
Cartridges: Kynoch Perfectly Gastight; Kynoch Patent Perfectly Gastight.
Example one: Kynoch Perfectly Gastight. Ga 12; Tc brown/black; De Kynoch's (On its own) lion's head Trade Mark; Kn All the tube printing is Kynoch; Bl (Double head) 12 + 13; Pr -CI; St E. J. Brown № 12 Rotherham; Wd white card/black with shot size only (cluster type) 6; Or British.

BUCKMASTER & WOOD: Wokingham, Berks.
Cartridges: A cartridge has been seen in 12-bore. It had the St KB. Or British. No other details were recorded.

BUGLERS: Ashford, Kent.
Cartridges: The National.
Example: The National. Ga 12; Tc pale greyish green/black; De Crested with two flags; Bl 8; Pr -CI; St NGN; Or British.

BULLEN BROTHERS: Truro, Cornwall.
Business: Ironmongers.
Cartridges:
Example: Unnamed. Ga 12; Tc light brown/not recorded; De Running rabbit; Kn This cartridge is by The Cogscultze Company; Or British. No other details recorded.

W. BUNTING: Cromford, Derbyshire.
Cartridges: Mullerite Green Seal.
Example: Mullerite Green Seal. Ga 12; Tc grey/black; Kn The tube printing is Mullerite with the extra printing, W. Bunting. Cromford. British loaded. Smokeless; Bl 10; Pr LCI; St SS; Wd red/black Special · Smokeless · 5; Or Not known, possibly British.

W. BURGESS: Malvern Wells, Worcs.
Cartridges:
Example: Unnamed. Pinfire. Ga 16; Tc brown quality/nil; Kn This cartridge was rimless with a double-length pin; Bl Not recorded; St W. Burgess. Malvern Wells. 16; Or Not known.

J. BURROW: 46 Fishgate, Preston. Also at Lowther St, Carlisle.
Business: Gunmaker.
Cartridges: The Economic.
Example: The Economic. Ga 12; Tc brown/silver; De A shield outline that contains all the wording; Kn Nitro Smokeless; Bl 7; Pr --I; St Burrow. Preston & Carlisle. Eley; Or British.

BUTCHER: Watton, Norfolk.
Cartridges: Butcher's name has been seen on an over-shot wad, white card/red.

BUTLER: Bampton, Devon.
Business: Cycle agent and ironmonger.
I have been told that their name has been seen on a cartridge.

BUYERS ASSOCIATION: 72 Wigmore St, London. W.
Cartridges:
Example: Unnamed. Ga 12; Tc blue/print colour not recorded; De A seal-like monogram with the letters B A; Bl 8; Pr -CI; St Buyers Association. London · № 12 Eley; Or British.

J. CALVERT: Walsden, Near Todmorden, Yorks.
Cartridges: Ejector.
Example: Ejector. Ga 12; Two-piece brass case with a light brown inner paper tube; Bl 10 + 57; Pr MCI; St J. Calvert. Walsden · № 12 Eley; Or British.

R. CAMBELL & SONS: Leyburn, Yorks.
Cartridges: The Wensledale.
Example: The Wensledale. Ga 12; Tc red/black; Kn Specially loaded by Cambell & Sons; Bl 8; Pr Not recorded; St R. Cambell & Son. No 12 Leyburn; Wd Brownish orange/black. Kynoch · Smokeless · 6; Or British.

Note the word 'Sons' on the tube printing and 'Son' on the headstamping.

CAMBRIDGE & CO: Carrickfergus, Northern Ireland.
Cartridges: The Antrim; The County Down; The Ulster.
Example: The County Down. Ga 12; Tc tan/black; De A snipe standing, R; Kn Smokeless cartridge; Pr, -CI; St EL; Wd Schultze* 8; Or British. No other details recorded.

W. CAMERON & CO: Ballymena, Northern Ireland.
Business: Gun and ammunition merchants.
Cartridges: The Cameronia; Cameron's Special.
Example: The Cameronia. Ga 12; Tc Greyish green/black; Kn All British product. The wording: 'If you miss don't blame this cartridge.' Specially loaded for W. Cameron & Co. Gun and ammunition merchants; Bl 8; Pr MCI; St ICI; Wd yellow/black. Special · Smokeless · 5; Or British.

CARR BROTHERS: Huddersfield, Yorks.
Cartridges: Ejector.
Example: Ejector. Ga 12; Two piece brass case with a brown inner paper tube; Bl (Double head) 10 + 57; Br MCI; St Carr Bros № 12 Huddersfield. Eley; Or British.

E. P. CARR & CO: Lower Parliament St, Nottingham, Notts.
Business: Dealers.
Cartridges: Kynoch's C. B. Cartridge Case.
Example: Kynoch's C. B. Cartridge Case. Ga 20; Tc deep red/black; De Kynoch's encircled lion Trade Mark; Kn Kynoch's case printing. For Nitro powders; Bl 7; Pr MCI; St KB; Wd White card/pink. Carr. Nottingham. 5; Or British.

W. C. CARSWELL: 4a Chapel St, Liverpool, Lancs (Merseyside).
Business: Gunmaker.
Cartridges: The Banshee; Carswell's Special.
Example one: The Banshee. Ga 12; Tc brown tan/black; De Coat of arms; Kn Incorporating the business of E. & G. Higham and Hooton & Jones; Bl 9; Pr MCI; St Kynoch 12 12 Birmingham; Wd Pale brown card/black. W. C. Carswell. Liverpool. (Inner) Smokeless. 5; Or British.
Example two: Carswell's Special. Ga 12; Tc Middle blue/black or dark blue; De and Kn As on example one; Bl 16; Pr MCI; St EN; Wd Similar to example one; Or British.

CARTRIDGE SYNDICATE: 20-23 Holborn, London. E.C.1.
Business: Cartridge sales.
Cartridges: Spartan; Spartan Deep Shell. May have sold The London.
Example: Spartan Deep Shell. Ga 12; Tc grey/black; Kn Loaded with a British Smokeless Powder; Bl 16; Pr LCI; St Made in 12 12 England; Wd red/black. British Shot 5; Or British.

Cartridges were loaded for them by The Trent Gun & Cartridge Works of Grimsby. The name of Cartridge Syndicate was not printed on the cartridges or their boxes.

F. G. CASSWELL: Midsomer Norton & Radstock, Somerset (Avon).
Business: Ironmonger.
Cartridges:
Example: Unnamed. Ga 12; Tc dark red/black; Kn Loaded with Amberite gunpowder; Bl 8; Pr MCI; St EL; Or British.

A similar cartridge has been seen loaded into a Remington case.

HERBERT CAWDRON: The Butlands, Wells Next The Sea, Norfolk.
Cartridges: The Holkham.
Example: The Holkham. Ga 12; Tc purple/black; Bl 16; Pr MCI; St KB; Or British.

A. CHAMBERLAIN: 18 Queen St, Salisbury, Wilts.
Business: Gunmaker.
Cartridges: The A. C. County; The A. C. Wiltshire; The Command; Pegamoid; The Sarum; The Stonehenge; The Wessex. Also Pinfires.
Example: The A. C. County. Ga 16; Tc red/black; Kn Specially loaded by A. Chamberlain. Telegrams Chamberlain, Salisbury.; Bl 7; Pr MCI; St EL; Wd yellow/black. Chamberlain. Salisbury. 5; Or British.

E. CHAMBERLAIN: Andover & Basingstoke, Hants.
Business: Gunmaker.
Cartridges: The Smokeless; The Universal.
Example: Unnamed. Ga 12; Tc olive green/black; De A shield. This has scrolled sides and the letter A at the top. Inside is a male lion standing in front of a tree; Kn Practical gunmaker, Andover & Basingstoke; Bl 16; Pr -IC; St Eley N.I. · № 12 · London; Or British.

SEPTIMUS CHAMBERS: Bristol, Cardiff & Shepton Mallet.
Business: Gunmaker.
Cartridges: Special; Patent No 15848.
Example: (Patent No 15848). Ga 12; Tc white/two-colour printing being red and dark blue; Kn This cartridge has three coloured sections. The top is white paper tube, the centre is red and the lower section tight to the base is dark blue. All the wording runs around the case in the white section and is as follows, (blue) Septimus Chambers, (red) (Patent No 15848), on the next line and still in red, Gunmaker, (blue) next line, Bristol, Cardiff & Shepton Mallet. The tube was printed before the case was constructed; Bl 10; Pr MCI; St Chambers № 12 Bristol & Cardiff. Kynoch; Or British.

B. E. CHAPLIN: 6 Southgate St, Winchester, Hants.
Business: Gunmaker.
Cartridges: The Ideal; The Winton.
Example: The Winton. Ga 12; Tc pink/black; De Trade Mark; Bl 8; Pr MCI; St ICI; Wd red/black. Precision Loading. 6; Or British.
B. E. Chaplin took over the business of Howard A. Davies; he also marketed a cartridge with the name 'Winton'.

S. CHITTY: 6 Lion St, Chichester, Sussex.
Cartridges: The Chichester Cross; The Wonder.
Example: The Wonder. Ga 12; Tc buttercup yellow/black; Kn Phone 428; Bl 6; Pr LCI; St SS; Wd White card/red. Patstone. Southampton. 6; Or Foreign.
This cartridge was loaded by Patstone's for Mr Chitty.

E. J. CHURCHILL: Leicester Square, London. W.C.2.
Business: Gunmaker.
Cartridges: 8 – Star; A. G. (Accuracy Guaranteed); Ejector; Express Cartridge XXV; The Field; The Imperial; The Pheasant; The Premier; The Prodigy; Special; Special Trapshooting Cartridge; Utility; The Waterproof Metal Lined.
Example one: 8 – Star. Ga 12; Tc vermilion red/silver; De An eight-pointed star with the figure 8 in its centre; Kn Hand loaded. Waterproof; Bl 8; Pr Medium brass cap being copper coated and fitted in an insert cup; St Non Corrosive 12 12 Smokeless; Wd orange/black. Churchill Gunmaker. London. 6; Or Foreign.
Note the name 8 – Star was not printed on the cartridge. A pre-war advertisement referred to it as 'Eight good points'. These were listed as the cap, the cartridge rim, brass head and metal lining, the paper tube, the powder, the shot charge, the wadding, the crimp or turnover. When marketed in 1926 they were priced at 16/- per 100.
Example two: The Field. Ga 12; Tc crimson/black; Kn Hand loaded at Orange Street Gun Works, London. W.C.2.; Bl 16; Pr MCI; St ICI; Wd orange/black. Churchill Gunmaker. London. 5; Or British.

CHAS CLARKE: 17 Winchester St, Salisbury, Wilts. Also London.
Cartridges: The Original J.W.G.
Example: The Original J.W.G. Ga 12; Tc pale greyish green/dark blue; Kn Sole manufacturer, Chas Clarke. 17 Winchester Street, Salisbury and London; Bl 8; Pr MCI; St Clarke № 12 Salisbury. Wd mauve/black. Amberite * * * 5½; Or Not known.

FRANK CLARKE: Castle St, Thetford, Norfolk.
Business: Ironmonger.
Cartridges: The Grafton; The Invincible.
Example: The Invincible. Ga 12; Tc pale greyish green/black; Bl 8; Pr MCI; St ICI; Or British.

When the firm of Bond were in business in Thetford they also loaded a cartridge named the Invincible.

H. CLARKE & SONS: Leicester, Leics.
Business: Gunmakers.
Cartridges: The Alma; The Express; The Midland.
Example: The Midland. Ga 12; Tc dark red/black; De The centre of the Leicester coat of arms with the wording on the scroll below it, 'Semper Eadem'; Square turnover; Bl 9; Pr MCI; St (Outer) H. Clarke & Sons № 12 Leicester. (Inner) Express Cartridge; Wd white card/red. Clarke · Leicester · 5; Or Not known.

CLARKE & DYKE: Salisbury, Wilts. Also, Southampton, Hants.
Cartridges: The J.W.G.; The Salisbury.
Example: The Salisbury. Ga 12; Tc dull orange/black; St Clarke & Dyke № 12 Southampton. No other details recorded.

CLATWORTHY: Taunton, Somerset.
Business: Ironmonger.
Cartridges: Ejector.
Example: Ejector. Ga 12; Two-piece brass case with an inner paper tube; St; Purdey stamping; Wd white card/red with the print forming a background for white wording. Clatworthy. Taunton. No other details recorded.

CLATWORTHY COOKE & CO: Taunton, Somerset.
Business: Ironmongers.
Cartridges: The Pheasant Cartridge.
Example: The Pheasant Cartridge. Ga 12; Tc middle blue/dark blue; De Small cock pheasant in scrub with his short tail horizontal, L; Kn Specially loaded by Clatworthy Cooke & Co Ld; Bl 8; Pr MCI; St ICI; Wd brown card/black. Eley-Kynoch Loaded 5; Or British.

Note that the tube printing shows a Clatworthy Cooke & Co's loading, while the over-shot wad shows an Eley-Kynoch loading. My guess is that it was an Eley-Kynoch load.

F. K. CLISBY: Marlow, Bucks.
Business: Gunmaker and cartridge-loading expert.
Cartridges: Special Loading.
Example: Special Loading. Ga 12; Tc orange/black; De A bird looking much like a snipe standing upright on a small clump, L; Bl 8. No other details recorded.

THOS CLOUGH & SON: King's Lynn, Norfolk.
Cartridges: The Sandringham.
Example: The Sandringham. Ga 12; Tc orange/black; Kn Specially loaded by T. Clough & Son, King's Lynn. Telephone, Lynn 2322; Bl 8; Pr MCI; St ICI; Wd white card/pencil markings, SSG; Or British.

THE CLUB CARTRIDGE CO: 2 Pickering Place, London. S.W.
Cartridges:
Example: Unnamed. Ga 12; Tc grey/blue; De A large round crest with the wording, 'Clubs are trumps'; Bl 8. No other details recorded.

The illustrated cartridge had a light grey paper case with Navy blue print.

COGSCHULTZE AMMUNITION & POWDER CO: London.
Business: Cartridge loaders.
Cartridges: The Bono; The Farmo; Gastight; The Molto; The Pluvoid; The Ranger; The Westro.
Example: The Westro. Ga 12; Tc dull orange/black; De A black oval crest forming a background for the wording Cogschultze London. Ammunition & Powder Co Ltd; Kn The Westro. Regd. Loaded Schultze. Made in Great Britain. Cogschultze Co's loading; Bl 8; Steel liner; Pr MCI; St Cogschultze 12 12 London; Or British.

This firm was founded with £10,000 in August 1911 and used Cogswell & Harrison's cartridge cases and the Schultze Powder Co's gunpowders. It ran for two years and it is possible that it may still have been in business at the outbreak of the First World War.

COGSWELL & HARRISON: New Bond St, The Strand. Later at Piccadilly, London.
Business: Gun and cartridge makers.
Cartridges: The Ardit; The Avant-Tout; The Blagdon; The Blagdonette; The Certus; Kynoch's Grouse Ejector; The Huntic; The Kelor; The Konkor; The Konor; The Markor; The Markoroid; The Midget; Nitro; Pegamoid; The Swiftsure; The Victor; The Victor Universal; The Victoroid.
Example one: The Markoroid. Ga 12; Tc orange tan/black; Kn The name of the cartridge is printed diagonally in longhand. Redg. All wording running around the tube with the names Cogswell & Harrison being printed as normal and inverted; Bl 9; Pr MCI; St Cogswell & Harrison 12 12 London; Wd orange/black. Vicmos Powder 5½; Or British.
Example two: The Victor Universal Standard Cartridge. Ga 20; Tc yellow/black; De The Cogswell & Harrison toothed Trade Mark; Bl 14; Pr LCI; St Cogswell & Harrison · 20 · ; Wd white card/black. Cogswell & Harrison * 5; Or British.

Founded by Benjamin Cogswell in 1770. The partnership with Edwin Harrison came in 1837. In 1932 the firm became liquidated but a new firm emerged in 1933 and continued until 1982. Their zigzag Trade Mark markings were registered in 1897.

JOHN COLBY EVANS: 4–5 Dark Gate, Carmarthen (Dyfed).
Business: Ironmonger.
Cartridges:
Example: Unnamed. Ga 12; Tc orange/black; De The top half of a sitting dog that has a duck in its mouth, R; Kn Specially loaded for John Colby Evans. General ironmongers. Ammunition dealers; Bl 8; Pr MCI; St EN; Wd white card/black. Smokeless·Diamond· 3; Or British.

F. J. COLE: 171 Cricklade St. Later at 26 Castle St, Cirencester, Glos.
Business: Gunmakers.
Cartridges: The Castle; The Champion; The County Favourite.
Example: The Champion. Ga 12; Tc middle blue/dark blue; Kn Specially loaded by F. J. Cole. Gunmaker, Cirencester. Telephone, Cirencester 40; Bl 8; Pr MCI; St ICI; Or British.

COLE & SON: Devizes, Portsmouth, Chippenham & Windsor.
Business: Gunmakers.
Cartridges: The Crown; The King Cole; The Signature.
Example: The Signature. Ga 12; Tc greyish blue/black; De Cole & Son written in longhand as a signature; Bl 16; Pr MCI; St EL; Wd white card/pink. Cole & Son · Devizes · 5; Or British.

It is not certain if this firm was at Windsor while they had a branch at Chippenham.

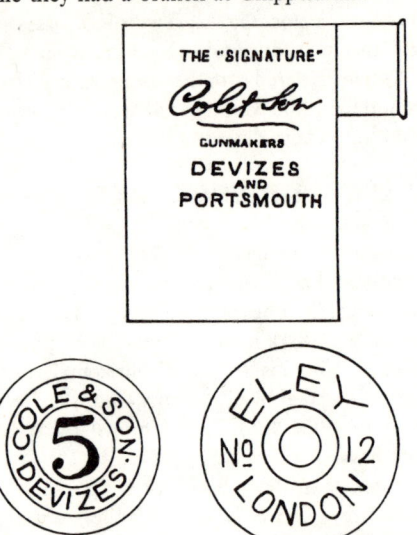

The Signature. Case colour: Jamaica blue with black print. Wad colour as illustrated: White card with pink print.

J. COLLIS: Strood, Rochester & Gravesend, Kent.
Cartridges: The All Round; The Famous Nulli-secundus. Also spelt, Nulli Secundus.
Example: The All Round. Ga 16; Tc firtree green/black; Kn Smokeless cartridge loaded by J. Collis Ltd; Bl 7 copper coated; Pr MBI; St Smokeless 16 16 Gastight; Or Foreign.

COLTMAN & CO: 49 Station St, Burton-upon-Trent, Staffs.
Business: Gunmakers and cartridge loaders.
Cartridges: The Burton; Ejector; The Field; The K.C. (Keepers Cartridge), also known as The Staffordshire Knot; The Marvel; The Partridge; Pegamoid; The Pheasant; The Rabbit.
Coltman's also loaded two special brands for a Wolverhampton firm. They were The Governor; The Victor.
Example one: The K.C. Ga 16; Tc black/gold; De A picture of the Staffordshire knot with the letters K and C and the wording 'Keepers Cartridge' written within it; Kn Gastight. High Velocity Cartridge. Loaded by Coltman & Co. Burton-upon-Trent; Bl 10; Pr MCI; St (Outer) Smokeless 16 16 Gastight. (Inner) R.W.S. Nuremberg; Or German.
Example two: The Governor. Ga 12; Tc kingfisher blue/black; De A Trade Mark picturing a ball-type centrifugal governor with the initials J. S. & S.; Kn Gastight High Velocity Cartridge; Bl 16; Pr MCI; St Leon Beaux & Co. 12 12 Milano; Wd orange/black with the print forming a background for the orange wording. Coltman · Burton · 6; Or Italian.

ROY CONWAY: Grimsby, Lincs (Humberside).
Cartridges: Ejector.
Example: Ejector. Ga 14; An all-brass case; St Kynoch № 14 Patent; Wd red/black. Name on wad only; Or British. No other details recorded.

ARTHUR CONYERS: Blandford Forum, Dorset.
Business: Gunmaker.
Cartridges: The Dorset County; The Express; The Express De Lux; Water Resisting Case.
Example: The Express. Ga 12; Tc's white/black or dark blue, yellow/black, orange/black; De Rabbit running and bounding over a young foxglove plant, FL; Kn Loaded and guaranteed by Arthur Conyers. Telephone No 7; Bl 8; Pr MCI; St ICI; Wd vermilion red/black with the print forming a background for the outer wording. Conyers · Blandford · (Inner) Special Loading 8; Or British.

CONYERS & SONS: Driffield, Pocklington & Blandford.
Business: Gunmakers.
Cartridges: The Express.
Example: The Express. Ga 12; Tc cream white/black; De Rabbit running and bounding over a young foxglove plant, Fl; Bl 8; Pr MCI; St EL; Or British.

C. COOK & CO: Bazzar, Leigh (County not known).
Cartridges: Smokeless Cartridge.
Example: Smokeless Cartridge. Ga 12; Tc ruby red/black; De Cock pheasant standing in grass with tail outstretched, L; Kn Specially loaded by C Cook & Co Ltd; Bl 8; Pr MCI; St KB; Or British.

COOMBES: address unknown.
Cartridges: Coombes' Champion.
Example: Coombes' Champion. Ga 12; Tc cream white/dark blue; Kn Specially Eley loaded; Bl 8; Pr MCI; St ICI; Or British. No other details recorded.

WILLIAM COOMBES: Frome, Somerset.
Cartridges: The Eclipse.
Example: The Eclipse. Ga 12; Tc light red/black; De A coat of arms in an oval; Kn Specially loaded. Kynoch's Patent. Perfectly Gastight; Bl (Double head) total 25; Pr MCI; St W. Coombes. Frome. Kynoch No 12; Or British.

GEORGE COONEY: Kells, Antrim or Kilkenny, Republic of Ireland.
Business: Hardware merchant.
Cartridges:
Example: Unnamed. Ga 12; Tc light brown/black; Kn Kynoch's Perfectly Gastight; Bl 25; Pr MCI; St KB; Wd red/black. K. S. G. Powder. BB; Or British. No other details recorded.

S. L. CORDEN: Warminster, Wilts.
Cartridges: The Quickfire.
Example: The Quickfire. Ga 12; Tc red/black; Bl 8. No other details recorded.

JOHN CORNISH: Okehampton, Devon.
Cartridges: The Oakment.
Example: The Oakment. Ga 12; Tc maroon/silver; De Rabbit bounding over grass, L; Kn Special Smokeless. British hand loaded; Bl 8; Pr MCI; St SS; Or Foreign.

CORNWALL CARTRIDGE WORKS: Liskeard, Cornwall.
Business: Cartridge loading and sales.
Cartridges: The Cornubia; The Cornwall; The Tamar; The Trelawney; The British Challenge.
Example: The Cornwall. Ga 12; Tc eau-de-nil/dark blue; De Cock pheasant standing in grass, R; Kn Smokeless. Hand loaded; Bl 8; Pr MCI; St ICI; Wd white card/black. Loaded in Cornwall. 7; Or British.

G. COSTER & SON: 145 West Nile St, Glasgow, Lanarkshire.
Business: Gunmakers.
Cartridges: G. C. & S.
Example: G. C. & S. Ga 12; Tc orange/black; Bl 8; Pr MCI; St EN; Or British. No other details recorded.

WALTER COTON: Coventry, Warwickshire (Midlands).
Cartridges: keepers.
Example: Unnamed. Ga 12; Tc yellow/black; De Cock pheasant standing in grass with tail horizontal, L; Bl 16; Pr MCI; St KN; Or British.

Walter Coton marketed various yellow/black-cased cartridges with a cock pheasant on the tube similar to the example above. These came in many different cases with variations in the Bl. I have seen cartridges in 2 inch, 2½ inch and 3 inch case lengths. The firm was blitzed during the Second World War. *Above right*: Case colour: Cowslip yellow with black print.

THE COUNTRY GENTLEMEN'S ASSOCIATION: address unknown.
Cartridges: The C.G.A. Improved Gastight Cartridge; The C.G.A. Improved Waterproof Cartridge; The C.G.A. Keepers Cartridge.

Example: The C.G.A. Keepers Cartridge. Ga 12; Tc pale green/black; De A crest; Kn By Kynoch Ltd; Bl 8; Pr MCI; St KB; Or British.

The C.G.A. Waterproof Cartridge (*opposite below*). Case colour: Nut brown with black print.

COX & CLARKE: 28 High St, Southampton, Hants.
Business: Gunmakers.
Cartridges: Fourten; The Southampton.
Example: Fourten. Ga ·410; Tc firtree green/black; Kn 28 High Street; Bl 8; Pr MCI; St Kynoch ·410; Wd white card/black. Shot size only (Cluster type) 8; Or British.

COX & MACPHERSON: Southampton, Hants.
Business: Gunmakers.
Cartridges: Special.
Example: Special. Ga 16; Tc light brown/blue; Bl 7; Pr MCI; St EL; Wd white card/pink. Cox & Macpherson. Southampton. 5; Or British.

COX & SON: 28 High St, Southampton, Hants.
Business: Gunmakers.
Cartridges: J.W.G. Special Cartridge; The Popular; The Southampton; The Star Special.
Example: J.W.G. Special Cartridge. Ga 12; Tc blue/black; Kn Supplied only by Cox & Son. Schultze powder; Bl 8; Pr MCI; St EN; Or British.

G. CREIGHTON: 8 Warwick Rd, Carlisle, Cumberland (Cumbria).
Business: Gunmaker.
Cartridges:
Example: Unnamed. Ga 12; Tc brownish orange/black; De The EBL Shield Registered Trade Mark with black letters and rope-type edging; Bl 8; Pr MCI; St EL; Wd white card/pink. G. Creighton. Carlisle. 5; Or British.

D. CROCKART & CO: Stirling, Stirlingshire.
Business: Gunmakers.
Cartridges:
Example: Unnamed. Ga 12; Tc brown/black; De A diamond-shaped crest with most of the wording within. Penetration. Pattern. Uniformity. Quality. Gastight. Metal lined; Bl 16; Pr MCI; St Eley Nº 12 Gastight; Wd white card/red. Red Star · Smokeless · 6½; Or British.

D. B. CROCKART: Perth, Perthshire.
Business: Gunmaker.
Cartridges: The Perth; The Spotfinder.
Example: The Spotfinder. Ga 20; Tc purple/silver; Kn Loaded by D. B. Crockart; Bl 7; Pr MCI; St ICI; Wd white card/black. Smokeless · Diamond · 5; Or British.

CROCKART & SON: Blairgowrie, Perthshire
Business: Gunmakers.
Cartridges: J.C. & S.
Example: J.C. & S. Ga 12; Tc orange/black; De A triangle with the initials, J.C. & S. within; Kn Specially loaded by Crockart & Son; Bl 8; Pr MCI; St ICI; Or British.

CROSS BROTHERS: 3–4 St Mary's St, Cardiff, Glam.
Cartridges: No cartridge information available.

R. CROWE. LTD: 63 Maldon Road, Great Baddow, Essex.
Business: Gun and sports stores.
The Crow. Case colour: Signal red paper case with black print. This is a Hull Cartridge Company load in an Italian case made by Fiocchi.

This cartridge has also been loaded in 16- and 20-bore Fiocchi cases with plastic tubes. All these cartridges are star crimp closure.

CURTIS'S & HARVEY'S: 3 Gracechurch St, London. E.C.
Business: Gunpowder and cartridge manufacturers.
Cartridges: Amberite; Feather Weight; Lined Nitro; The Marvel; Ruby; Smokeless Diamond; Unlined Nitro.
Example: The Marvel. Ga 12; Tc pale greyish green/black; Kn Smokeless cartridge; Bl 8; Pr MCI; St Curtis's & Harvey. L*d*. London. № 12 (Small E.B.L. shield); Wd white card/black. Hounslow Loading · C&H · Shot size, B; Or British.

Most of their cartridges were coloured blue/black, brown/black or grey/black. They often took their name from the type of powder loaded. C & H eventually amalgamated with the powder firm John Hall, who had mills at Faversham in Kent. In November 1918 they themselves merged into Explosives Traders Limited.

J. H. CUTTS: Macclesfield, Cheshire.
Cartridges: The Special.
Example: The Special. Ga 12; Tc pink/black; Bl 8; Pr MCI; St Eley № 12 Gastight; Or British.

T. DAINTITH: Warrington, Lancs (Cheshire).
Business: Gunmaker.
Cartridges: Eley Pegamoid.
Example: Eley Pegamoid. Ga 12; Tc brown/black; Kn Eley gas-tight waterproof cartridge case. Made in Great Britain from Pegamoid Brand paper. Specially loaded by T. Daintith, Gunmaker. Warrington; Bl 16; Steel lined; Pr MCI; St ELG; Or British.

N. S. DALL: Chichester, Sussex.
Cartridges: An old cartridge head has been found with their name and town on the stamping.

W. DARLOW: Bedford & Norwich.
Business: Gunmaker.
Cartridges: The Big Bag; The Lightning.
Example: The Lightning. Ga 12; Tc pale eau-de-nil/dark blue; De Cock pheasant standing in grass with tail horizontal, L; Bl 8; Pr-CI; St W. Darlow № 12 Bedford & Norwich Kynoch; Wd red/black W. Darlow · Bedford · 5½; Or British.

DARLOW & CO: 8 Orford Hill, Norwich, Norfolk.
Business: Gunmakers. Later, Gunsmiths.
Cartridges: The Big Bag; The Castle; The Lightning; The Orford.
Example: The Castle. Ga 12; Tc light grey/blue; Kn All British smokeless cartridge; Bl 8; Pr MCI; St ICI; Or British.

FRANCIS DAVIE: Elgin, Morayshire.
Cartridges: The Moray Cartridge.
Example: The Moray Cartridge. Ga 12; Tc orange/black; Kn Loaded by Francis Davie of Elgin; Bl 8; Pr MCI; St EN; Wd orange/black. F. Davie · Elgin · 5; Or British.

ALFRED DAVIS: 4 Bishopsgate Churchyard, Old Broad St, London. E.C.2.
Cartridges: The Bishopsgate.
Example: The Bishopsgate. Ga 12; Tc maroon/gold; Kn From J. Blanch & Son. Our special loading. Telephone, London Wall 1130; Bl 8; Pr MCI; St Special smokeless. Foreign Made Case; Wd white card/red. Special · Smokeless · 5; Or Foreign.

HOWARD A. DAVIS: 6 Southgate St, Winchester, Hants.
Business: Gunmaker.
Cartridges: The Flight; The Winton.
Example: Unnamed. Ga 12; Tc firtree green/black; De An alighting duck (see illustrations pages 25–28); Bl 8; Pr MCI; St ICI; Wd red/black. Precision+Loading+ 6; Or British.

This business was later taken over by B. E. Chaplin. He also continued to market a Winton Cartridge.

M. H. DAVIS & SONS: Aberystwyth, Cardigan (Dyfed).
Business: Ironmongers.
Cartridges: Details of their cartridges are not known.

Following pages:
1 *Some colourful pre-war examples*
2 *A selection of Australian cartridge boxes*
3 *Note several deep-brass cases in this display*

Top row	BRITISH HAND LOADED / JOHN LANGDON / GUNSMITH / 20 ST MARY ST, / TRURO / PHONE 907
	BRITISH HAND LOADED / FUSSELL'S / GUN-MAKERS / 118-119 CHEAPSIDE, LONDON, E.C.2. / 55 CROSS ST., ABERGAVENNY / 81 STATION ROAD, PORT TALBOT / 2 DOCK STREET, NEWPORT, MON. / 65 HIGH ST., NEWPORT, MON.
	LOADED & GUARANTEED BY / ARTHUR CONYERS / BLANDFORD / CARTRIDGE
	THE "DORSET" COUNTY
	"PNEUMATIC" NO. 1 / WATERPROOF / METAL LINED / Trade "PNEUMA PNEUMATIC" Marks
	GRANT & LANG, / CHARLES LANCASTER & CO. / WATSON BROTHERS / "ROCKETER" / MADE IN GREAT BRITAIN
Middle row	PRIMROSE SMOKELESS
	PAGE-WOODS' / ANTI-RECOIL / PATENT / CARTRIDGES / THE
	BARTRAM'S / "HARD HITTERS" / (SMOKELESS POWDER) / Made in Great Britain
	ELEY / "ROCKET" / Cartridge / SHEWS THE FLIGHT OF / in Great Britain
	EARLE'S / Special Cartridge / TELEPHONE NO. 2164 / BRIDGNORTH
Bottom row	THE "HUMBER" / CARTRIDGE / LOADED BY / R. ROBINSON / (GUNMAKERS) LTD. / 7 QUEEN STREET
	THE / ALBION / LOADED WITH / LIGHTNING / POWDER / THE SCHULTZE Co. Ld.
	LOADED BY / H. G. JACKSON / IRONMONGER & GUNSMITH / BUNGAY & HALESWORTH
	YELLOW SEAL / mullerite / SMOKELESS
	PRIMAX / CARTRIDGE / KYNOCH LOADED / SMOKELESS

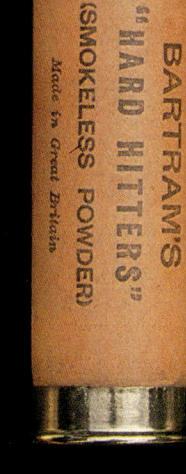

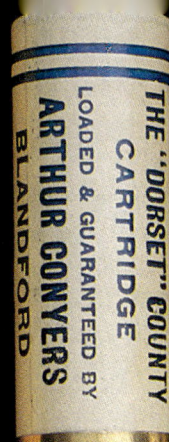

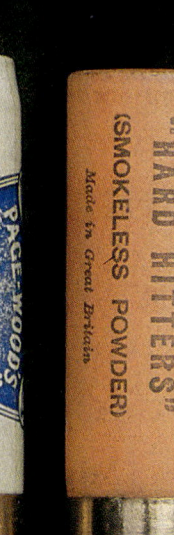

WINCHESTER
QUAIL LOAD
25 SHOTGUN SHELLS
WARNING KEEP OUT OF REACH OF CHILDREN
CLUB SPECIAL
WINCHESTER
TARGET LOAD
25 SHOTSHELLS
WARNING: Keep out of reach of children.
ELEY
TRAPSHOOTING
ICI
NON-CORROSI
PRIMEF
BLUE STAR
NON-CORROSIVE PRIMER
ICI
SMOKELESS
12 GAUGE
STAR CRIMPED
REMINGTON-PETERS
DOVE & QUAIL LOADS
PLASTIC SHELLS
25 SHOTGUN SHELLS
WARNING • KEEP OUT OF REACH OF CHILDREN
MADE IN U.S.A.
BLUE STAR
IMI
LONG RANGE
HIGH VELOCITY
GAME LOAD
CAC
CHAMPIONSHIP
SHOTGUN CARTRIDGES
with
NON-CORROSIVE
PRIMER
GRAND PRIX
NON-CORROSIVE PRIMER
ICI
SMOKELESS
12 GAUGE
STAR CRIMPED
WINCHESTER
AUSTRAL
NATIONAL CHAMPIONS
12 GAUGE CARTRID
GRIFFITH
WINCHESTER
LIGHT GAME
HIGH VELOCITY SHOTGUN SHELLS
MAGNUM
ICI
WEATHER
PROOFED
ICIL
Special
ICI
LONG RANGE WATERPROOFE

CAC
SMOKELESS
CLAY LOAD
12 GAUGE 6 2½ SHELL

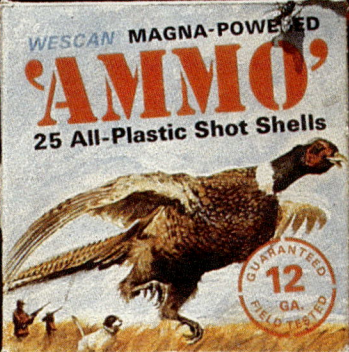

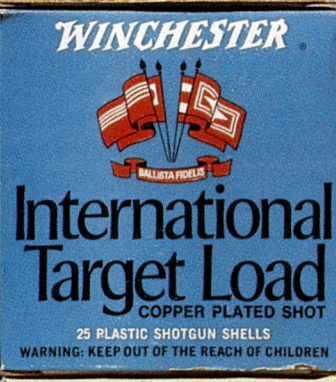

WESCAN MAGNA-POWERED
'AMMO'
25 All-Plastic Shot Shells
GUARANTEED FIELD TESTED
12 GA.

WINCHESTER
HIGH VELOCITY DUCK LOAD
25 SHOTGUN SHELLS
WARNING KEEP OUT OF REACH OF CHILDREN

CUSTOM LOADED
Geo. Biggs
SHOTSHELLS

WINCHESTER
BALLISTA FIDELIS
International Target Load
COPPER PLATED SHOT
25 PLASTIC SHOTGUN SHELLS
WARNING: KEEP OUT OF THE REACH OF CHILDREN

TARGET
ICI
SPECIAL TRAP & SKEET LOAD

WINCHESTER
HIGH VELOCITY RABBIT LOAD
25 SHOTGUN SHELLS
WARNING: KEEP OUT OF REACH OF CHILDREN

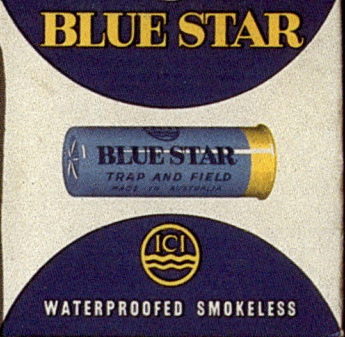

2¾ SHELL
LONG RANGE

BLUE STAR
BLUE STAR
TRAP AND FIELD
MADE IN AUSTRALIA
ICI
WATERPROOFED SMOKELESS

A.C.T.A
CARTRIDGES LOADED FOR
THE AUSTRALIAN CLAY TARGET ASSOCIATION
NATIONAL CHAMPIONSHIPS
1969
GEELONG, VIC.

CAC Plastic
LONG RANGE
LONG RANGE
SHOTGUN CARTRIDGES

IMI
BLUE STAR
SKEET
25 CARTRIDGES 12 GAUGE

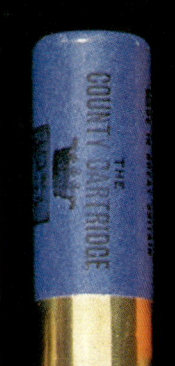

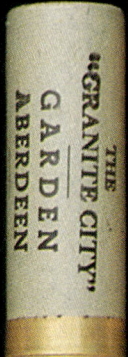

Row	Cartridges
1	"TIGER" BRAND LOADED BY ROBERT LISLE GUNMAKER DERBY · THE VICTORIA · (crossed flags) · THE LEICESTER · THE COUNTY CARTRIDGE · ADKIN'S "DEMON" BEDFORD
2	SNEEZUMS' "ANGLIA" · EVERY AN NITRO C'G'E THE AMBERITE · THE "GRANITE CITY" GARDEN ABERDEEN · R. H. ROWLAND WOODBRIDGE SPECIAL LOADING · "COTON'S" SPECIAL KEEPERS' CARTRIDGE · WEBBER'S "ISCA" Specially Loaded with Highest Grade Powder
3	"OVERSEAS" THE Little Fruit Cartridge · THE "EDWARD" SELECTED SMOKELESS POWDER · KNIGHTS · "SWIFT" · "HENRITE" · THE ANGLIAN HIGH VELOCITY
4	CHURCHILL "UTILITY" · BRADDELL'S TRADE MARK · "CLUB" Specially loaded by ALEX MARTIN GLASGOW, ABERDEEN · FRANK DYKE'S "SUPREME" GASTIGHT SINOXID · BEESLEY LONDON BY APPOINTMENT · NOBEL'S Sporting Ballistite CARTRIDGE

GEORGE H. DAW: 67 St James's St. Also at 57 Threadneedle St, London.
Business: Central fire cartridge developer.
Cartridges:
Example: Unnamed. Ga 12; Tc brown quality/nil; Bl 6 (In brass); Pr -CI; St (Reversed) G. H. Daw's № 12 Patent; Or British.
Mr Daw was one of the leading designers in the early development of the central fire cartridge. Daw's cartridges date from 1861 and in 1866 he took Colonel Boxer and Messrs Eley Brothers to court for what he considered were infringements made upon his patents. Eley Brothers won the case. Colonel Boxer designed a cartridge case with coiled brass constructed tube. This kind of constructed case has since often been referred to as Boxer Type.

ARTHUR DENNIS: Great Dunmow, Essex.
Cartridges: The Demon.

DESBOROUGH & SON: Derby, Derbyshire.
Cartridges: The Dovedale.
Example: The Dovedale. Ga 12; De A running hare; St EL; Or British. No other details recorded.

JOHN DICKSON & SON: 21 Frederick St, Edinburgh, Midlothian.
Business: Gun & rifle makers.
Cartridges: The Special Blue Shell; The Capital; The Ejector; Dickson's Favourite; The Jubilee.
Example: Dickson's Favourite. Ga 12; Tc crimson/black; De The Dickson's lion motif. This is a male lion standing on a large rock, L; Bl 16; Pr MCI; St ICI; Wd red/black. John Dickson & Son · Edinburgh · 6; Or, British.

John Dickson & Son incorporated the business's of Alex Henry, Alex Martin and Mortimer & Son.

DINWOODIE & NICHOLSON: Thornhill (County not known).
Cartridges: Eley's Best Gastight Case.
Example: Eley's Best Gastight Case. Ga 12; Tc blue/black; Kn Loaded with Curtis's & Harvey's Smokeless Diamond powder; Bl 16; Pr MCI; St ELG; Or British.

E. DISTIN & SON: Totnes, London.
Cartridges: Demon Cartridge.
Example: Demon Cartridge. Ga 12; Tc flesh pink/dark blue; Bl 10; Pr MCI; St Joyce № 12 London; Wd pale orange/black. Schultze · * · 5; Or British.

DIXON & CO: Aston Common, Birmingham, Warwickshire (W. Midlands).
Cartridges: Special; Special Hand-Loaded; Special Hand-Loaded Pigeon Cartridge.
Example: Special Hand-Loaded. Ga 12; Tc maroon/silver; Case length, 2¾ inch (70 mm); Kn Kynoch Patent Perfectly Gastight; Bl (Double head) 10 + 15; Pr MCI; St Dixon & Co № 12 Aston. Kynoch; Wd powder orange/black. Schultze 8; Or British.

DOBSON & ROSSON: Derby, Derbyshire.
Cartridges:
Example: Unnamed. Ga 12; Tc dark blue/nil; Bl 8; St Dobson & Rosson. Derby. No.12: No other details recorded. This cartridge came to light in New Zealand. It is quite possible that Charles Rosson or his father was in business with a Mr Dobson in Derby prior to the business of Rosson on its own.

W. G. DONALDSON: Grantown-on-Spey, Morayshire.
Cartridges: The Triumph.
Example: The Triumph. Ga 12; Tc pink/black; De A round Trade Mark with a hand holding up a sword; Kn Smokeless cartridge; Square turnover; Bl 16; Pr MCI; St EN; Wd red/black. Smokeless. 5½; Or British.

J. D. DOUGALL & SONS: Glasgow & London.
Business: Gun and rifle makers.
Cartridges: Pegamoid.
Example: Pegamoid. Ga 12; Tc brown/black; De The EBL Shield Registered Trade Mark with outlined letters and rope-style edging; Kn Eley's Water-proof. Gas-tight Cartridge Case. 'Pegamoid' Patent; Bl 11; Pr MCI; St J. D. Dougall & Sons. Glasgow · № 12 Eley; Wd white card/black. (Cluster type) 6; Or British.

This firm was established in 1760. During 1850, J. D. Dougall left Trowgate for Gordon Street in Glasgow. Here they remained until the turn of the century. Premises were also obtained at 59 St James's Street in Piccadilly, London. There the Glasgow origins were acknowledged.

DOWNING: Southwell, Notts.
Cartridges:
Example: The Schultze Co's Westminster. Ga 12; Tc pinkish orange/black; De The Schultze oval Trade Mark with the clenched fist with lightning. Kn Loaded with Schultze. Shultze Co's Loading; Bl 8; Pr MCI; St EL; Wd pink/black with the shot size hand stamped with a purple ink pad. Downing · Southwell · (Inner) Smokeless · Shot size 6 or 9; Or British. Name on wad only.

S. DUNCAN & SONS: 62 Albany Rd, Hull, Yorks (Humberside).
Business: Gunmakers.
Cartridges: Duncan's Special Load.
Example: Duncan's Special Load. Ga 12; Tc eau-de-nil/dark blue; Bl 8; Pr MCI; St ICI; Wd white card/nil. (Pencilled shot size) 2; Or British.

FRANK DYKE & CO: 10 Union St, London. S.E.1.
Business: Manufacturers and merchants.
Cartridges: Rabbit; Shamrock; Special; Special Gastight; Supreme; Yellow Wizard.
Example one: Frank Dyke's Supreme. Ga 12; Tc ice grey/black; De Cock pheasant standing on its left leg with the head turned towards the back. A wood forms the background, L; Kn Gastight. Sinoxid. British loaded. Waterproof; Bl 15; St SS; Wd yellow/black. Special Smokeless 6; Or Foreign.
Example two: Yellow Wizard. Ga 12; Tc buttercup yellow/black; Kn The name the Yellow Wizard, in large black capital letters. Sinoxid. Rustless priming. Smokeless. Loaded in England; Bl 7; Pr MBI; St SS; Wd yellow/black. Special · Smokeless · 5; Or Foreign.
Frank Dyke & Co Ltd. loaded very many cartridge brands for other firms. Most of his own brand cartridges never carried his name including the Yellow Wizard. In fact, the very first Yellow Wizards did not have any tube print. This firm lost its premises during World War Two due to the bombing.
Above right: The lower cartridge was discontinued with World War Two. Case colour: Deep fiesta yellow with black print. Wad colour: Yellow with black print.

A. H. DYKES: Stowmarket, Suffolk.
Cartridges:
Example: Unnamed or name not known. Ga 12; Tc maroon/silver; De Pheasant; Bl 8; St SS; Or Foreign. No other details recorded.

EARLE: Bridgnorth, Salop.
Business: Ironmonger.
Cartridges: Earle's Special Cartridge.
Example: Earle's Special Cartridge. Ga 12; Tc orange/black; Kn Telephone No, 2164; Bl 8; Pr MCI; St ICI; Wd yellow/black. Smokeless · Diamond · 6; Or British.

EASTMOND: Torrington, Devon.
Business: Ironmonger.
Cartridges: This firm once had their name printed on Mullerite cartridges. They were also known to have sold Eley Brothers and Curtis's & Harvey's brand cartridges.

THE 'E.C.' POWDER CO: 40 Broad St, London. E.C.
Business: Gunpowder manufacturers.
Cartridges: E.C.; E. C. Pegamoid.
Example: E. C. Ga 12; Tc brown/black; De The large round and sun-edged E.C. crest (It may have been a Trade Mark); Bl 16; Pr MCI; St ELG; Wd white card/red. E.C. 6; Or British.

Known as The Explosives Company, it was easily confused by name with other gunpowder companies, the Noble Explosives Company was one instance. To avoid this, the first letters of the company's name were used, so arriving with the initials, 'E.C.'. It is believed that the cartridge-loading side of this firm later became The New Explosives Company. Anyhow, both these firms were wound up in the new company of Explosives Trades Limited by the end of 1918.

R. E. EDMONDS: Stalham, Norfolk.
Business: Ironmonger.
Cartridges: The Stalham; The Stalham Superior.

EDNIE & KININMONTH: Forfar, Angus.
Business: Ironmongers and seed merchants.
Cartridges:
Example: Unnamed. Ga 12; Tc cream/black; De Running rabbit; Bl 9; Pr MCI; St EL; Or British. No other details recorded.

EDWARDS: Newport, Monmouth (Gwent).
Business: Gunsmith and cartridge maker.
Cartridges: The Newport.
Example: The Newport. Ga 12; Tc green/black; De Picture of a large goose or game bird; Kn Gunsmith & Cartridge maker; Bl Not recorded; Pr MCI; St EN; Or British.

C. G. EDWARDS & SON: Plymouth, Devon.
Business: Gunmakers.
Cartridges: The Eddystone; The Smeaton.
Example: The Eddystone. Ga 12; pale green/black; De The Eddystone Lighthouse; Bl 8; Pr MCI; St EL; Or British.

The Eddystone (*above right*). Case colour: Cream-yellow paper tube with blue-black print.
Wad colour as loaded in the example drawn: A mid-yellow card with no print. The cartridge was loaded with shot.

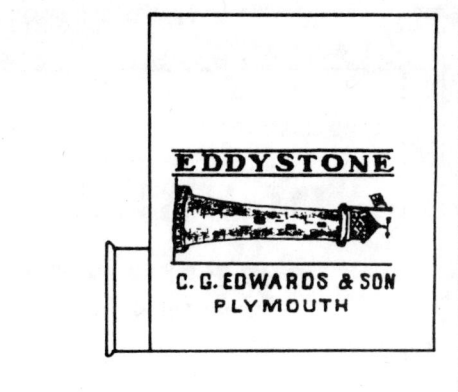

EDWARDS & MELHUISH: Harborne, Birmingham, Warwickshire (Midlands).
Cartridges: The Edmel.
Example: The Edmel. Ga 16; Tc crimson/black; 3-inch case length (75 mm); Kn Smokeless; Bl 16; Pr LCI; St Special 16 16 Gastight; Wd red/black. Special · Smokeless · Shot size, AA; Or Not known.

ELDERKIN & SON: Spalding, Lincs.
Business: Gun repair specialists.
Cartridges: The Premier.
Example: The Premier. Ga 12; Tc orange/black; Bl 8; Pr MCI; St ICI; Or British.

ELEY BROTHERS: 254 Gray's Inn Rd, London.
Business: Ammunition manufacturers. (Military and sporting): Needle-gun, Pinfire, Patent wire, Centre-fire.
Depots: Birmingham. 29 & 30 Whittal Street. (Eley Bros.)
Glasgow. 68 Mitchell Street. (James Ferrier & Son.)
Dublin. 9 Dawson Street. (Eley Bros.)
Exeter. Fore Street. (Evans Gadd & Co.)
Overseas depots. Belgium, Canada, United States of America, Italy with representatives in London, Australia, Cuba, Scandinavia, South Africa, South America and British North America.
Patent wire Cartridges. The Green Cartridge.

ELEY'S
DOUBLE WATERPROOF
CENTRAL-FIRE CAPS.

These Caps are now well known and approved, being found superior to all others for their certainty and rapidity of fire, either in dry or wet weather.

For India and the Colonies, or any climate where Caps may be exposed to great vicissitudes of heat, cold, or moisture, they are particularly recommended, as they cannot be injured by any amount of exposure to wet or heat, nor their qualities impaired, if kept for years in a tropical climate. The ignition at all times is safe and certain, whilst in humid weather the discharge is as instantaneous as with the ordinary Cap on the dryest day.

They have been much approved for the Rifle in Foreign Field Sports, where the Cap is often allowed to remain a long time upon the nipple.— See *General Jacob on Rifles and Projectiles*.

Being perfectly Waterproof, they will bear immersion in sea water for days without injury; but, when testing them in this manner, it is necessary to blow the water out of them before placing them upon the nipple.

Waterproof Caps made expressly for Duck Guns.

Waterproof Caps for ENFIELD and other RIFLES.

Metal-lined Anticorrosive Percussion Caps for Guns and Pistols.

Waterproof Copper Cartridges for Revolving Pistols.

Ditto   ditto   for Sharps' Pistols, &c.

ELEY'S **Bulleted Breech Caps** for Noiseless Pistols.

Eley's Sizes of Caps correspond with the Birmingham Sizes, as per annexed Table.											
ELEY'S	5	0	7	8	9	24	10	11	18	12	13	14
BIRMINGM	43	44	46	48	49	50	51 & 52	53 & 54	55 & 56	57	58	

Where there are two numbers of the BIRMINGHAM Sizes corresponding with only one of ELEY'S, it is in consequence of two numbers being of the *same size*, varying only in the LENGTH of the Caps.

TO BE HAD OF ALL RESPECTABLE GUNMAKERS.

Eley Brothers, MANUFACTURERS, **London**.

Colour, green. Kn Made for long-distance shooting.

The Royal Cartridge. Colour, red. Kn Intended for the second barrel but can be used in both.

The Universal Cartridge. Colour, yellow. Kn For use in first barrel, contains no wire.

Pinfire Cartridges. All information given here on gauges is taken from their 1902 price list.

Blue Best Quality. (Not listed in 1902.) Ga's seen are 12, 14, 16.

Brown Quality. Ga's 10, 12, 14, 16, 20, 24, 28.

Green Extra Quality. Ga's 8, 10, 12, 14, 16, 20, 24, 28, 32.

Red Quality. (Not listed in 1902.) Ga seen, 14.

Early centre fire cartridges. All information given here on gauges is taken from their 1902 price list.

Blue Best Quality (Not listed in 1902.) Ga's seen are 12, 16.

Brown Quality. Ga's 10, 12, 14, 16, 20, 24, 28.

Green Extra Quality. Printed in black the wording Eley's Gas-tight Cartridge Case; Ga's 4, 8, 10, 12, 14, 16, 20, 24, 28, 32. ·410, ·360. Note: Similar cartridges seen with Tc brown/black in many gauges (but not listed in 1902).

Early 1900's Cartridges. Ejector. (Paper cased cartridges with an outer brass seamless tube.) Ga's 8, 10, 12, 14, 16, 20, 24, 28, 32: Note, loaded with many gunpowders.

The E.B. Nitro Case. Ga's 12, 16, 20. Note: loaded with many gunpowders.

Improved Gas-Tight. Also known as The E.B.L. (With double heads and long shells.) Ga's 8, 10, 12, 14, 16, 20, 24, 28, 32, ·410.

Pegamoid. (The name 'Pegamoid' was an Eley Bros. patent for a waterproof type of brand paper.) All firms in this cartridge list with a brand 'Pegamoid' are made with this paper. Ga's 10, 12, 16, 20, 24, 28.

Solid Drawn Brass. Ga's 8, 10, 12, 14, 16, 20, 24, 28, 32. Thin brass, ·410.

Early Centre-fire Cartridge Cases for individual gunpowders. Hall's Smokeless Cannonite. Tc red; Ga's 12, 16, 20.

Amberite. Tc grey; Ga's, 10, 12, 16, 20, 24.

Cooppal Smokeless Game. Tc red; Ga's, 12, 16, 20.

Cooppal No, 2. Tc maroon; Ga's, 12, 16, 20.

E.C. Tc red; Ga's, 10, 12, 16, 20, 24.

Schultze. Tc buff; Ga's 10, 12, 16, 20, 24.

Shot Gun Riffleite. Tc red; Ga's 12, 16 20.

Sporting Ballistite. Tc primrose yellow; Ga's 12, 16, 20.

S.S. Smokeless. Tc grey; Ga's, 12, 16, 20.

1920 listed Short Smokeless Cartridges. Ga 12 only. Lancaster's Pygmies. 2¼ inch (56 mm) case length. The Parvo. 2 inch (50 mm) case length. Loaded with Sporting Ballistite.

Tom Thumb. 2 inch (50 mm) case length. Loaded with a Curtis's & Harvey's powder.

The Midget. 2¼ inch (56 mm) case length. Loaded with Cooppal No, 2.

Later brands. Centre-fire Cartridges. Achilles; Acme; Aquoid; Blacktwenty; Comet; D.S. Gastight Deep Shell Unlined; The ⅜" Shell, Gastight; E. B. Nitro; Ecar; Eley Smokeless; Ejector; Elite; Eloid; Erin-Go-Bragh; Fourten; Grand Prix; Juno; Lightmode; Eley's Lined Nitro Case; Mars; Neptune; Pegamoid; Pluto; Quail; Rocket; S.A.; Thor; Titan; Universal; V.C.; Vulcan; Zenith.

At the time the above brands were marketed, special cases were being manufactured for individual gunpowders. Also made were small batches of cartridges for breech-loading punt guns. One of these cartridges that I have seen carried the name Lightmode.

Example one: Gastight Cartridge Case. Ga 4; Tc dark green/black; Kn Eley's Gastight Cartridge Case; 4 inch (100 mm) case length; Reinforced paper around the base of the tube; Bl 13; Pr SCI; St EL; Wd white card/purple ink. Shot size only, BB; Or British.

Example two: Gastight Cartridge Case. Ga 14; Pinfire; Tc dark green/black; De Early type EBL Shield Registered Trade Mark; Kn Eley's Gastight Cartridge Case; Bl 10; St ELG; Or British.

Example three: S.A. Ga 16; Tc dark green/black; De EBL Shield Registered Trade Mark; Kn Eley loaded; Bl 15; Pr MCI; St EL; Or British.

Example four: Eley Gastight Cartridge Case for Amberite Smokeless Powder. Ga 12; Tc grey/black; De EBL Shield Registered Trade Mark with outlined letters and rope-type edging; Bl 16; Pr MCI; St ELG; Wd white card/red. Eleys · Loading · 4; Or British.

Example five: Pluto. Ga 12; Tc orange/black; De EBL Shield Registered Trade Mark with black letters and rope-type edging; Kn Eley loaded; Bl 8; Pr MCI; St EL; Wd white card/black. Eley × Loaded × 5; Or British.

(*Overleaf*): Eley Short Range Quail Cartridge: Case colour: Off-white with black print. Wad colour as illustrated: White card with red print.

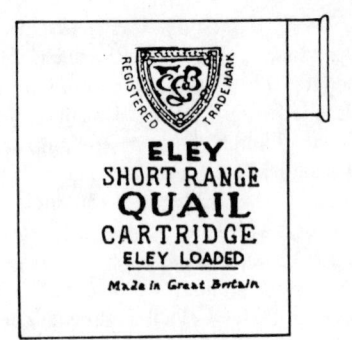

The early days of Eley Brothers are lost in obscurity though it is known the first two proprietors were brothers. Charles and William Eley were known to have been in business in 1828, but they suffered a major set-back when William was blown up by fulminating powder in the factory in 1841. At the time of the accident he was 46. Charles continued with the business and introduced his three sons – William Thomas, Charles and Henry. Together they ran the firm for 30 years.

Cartridge manufacture started in 1827 with the Eley Patent Wire Cartridges. These were constructed from a thin wire mesh which formed a cage containing the lead shot. The entire cartridge was then covered with a paper skin and placed in a paper wrapper. These cartridges were designed to be rammed home into muzzle-loaders and it was claimed they had several advantages over the conventional method of loading. The shot, for instance, did not have to be measured out in the field, there was less lead fouling in the barrels and the shot load, held together by the wire cage over a considerable distance in flight, produced tighter patterns. The patents for these cartridges were bought from a Frenchman, a M. Jenour. In 1837 Eley Brothers claimed another first when they marketed the very first waterproofed percussion caps.

Eley exhibited many of their products worldwide. They entered the Great Exhibition of 1851 and an exhibition in Vienna in 1873, where they were awarded prize medals for 'excellence of manufacture and materials'. They won the highest award in Chicago in 1893 and a gold medal in Antwerp the following year. In 1898 in Paris and again in 1900 they won the Grand Prix, subsequently giving this name to a range of their most successful cartridges. This dates Grand Prix to the start of the 20th century.

They not only sold their own brands but also retailed capped cases ready for loading and many other firms purchased from them. Firms nearly always bought the cases with the tubes ready printed and for a small extra charge could have their own names printed on the over-shot wads and stamped into the brass case.

Because they required more capital, Eley went public in March 1874. They continued to expand and marketed their cartridges in competition with firms such as George Kynoch & Co Ltd. After the end of the Great War many of the larger firms were merged into one known as the Explosive Trades Ltd. Eley Bros were one of the larger of the companies absorbed though they were not officially liquidated until 1928.

William was the first of the three brothers to die, and Henry resigned in 1902 shortly after the death of his other brother Charles. The name lives on, however, and Eley is synonymous with best British cartridges.

ELLICOTT: Cardiff, Glam.
Cartridges: The Ellicott.
Example: The Ellicott. Ga 12; Tc orange tan/black; Kn Gas-tight Cartridge Case for E.C. Gunpowder; Bl 12; St Ellicott. Cardiff. No other details recorded.

WILLIAM ELLICOTT: Broad St, Launceston, Cornwall.
Cartridges:
Example: Unnamed. Ga 12; Tc dark green/black; Kn Kynoch's Patent Perfectly Gas-tight Cartridge Case; Bl (Double head) 7 + 18; Pr MCI; St Ellicott № 12 Launceston; Wd brown card/black. Schultze o * o 4; Or British.

An advertisement for the cartridges of Ellicott of Launceston may be seen on the front of the very first '*Shooting Times*' publication dated the 9th September, 1882.

H. C. ELLIOTT: Lowfield St, Dartford, Kent.
Cartridges: The Smasher.
Example: The Smasher. Ga 12; Tc red/black; Bl 9; St It had Elliott's name on it. No other details recorded.

ELEY'S
PATENT WIRE CARTRIDGES,

For Shooting Game, Wild Fowl, &c., at long distances.

As there are few sportsmen who are not in the habit of using these Cartridges, they are so well known as to make a description of them scarcely requisite. The shot is packed within a wire cage, which is constructed so as to allow them to escape gradually while the charge is in motion. They cause all guns to shoot with double the strength which can be obtained by the ordinary mode of loading, and with much greater regularity, as each shot retains its spherical form.

Those in yellow cases, termed **Universal Cartridges**, are intended for the first barrel in the commencement of the season, or for wood shooting. These contain no wire.

The **Royal Cartridge** is intended for the second barrel at the commencement of the season, enabling the sportsman to take a double shot even when birds rise wildly. When the season is advanced, they may be used with success in both barrels.

The **Green Cartridge** is intended for wild fowl, and shooting at long distances. They are made for foreign field sports, with large mould shot, which is found very effective at large game, where the sportsman has not a rifle in the field.

** For Testimonials and Recommendations, see the following Authors on Shooting—Colonel HAWKER, Capt. R. LACY, T. B. JOHNSON, THOS. OAKLEIGH, "Nimrod," "The Old Sporting Magazine," W. WATT, M. MANGEOT, FORESTER'S "Field Sports," ST. JOHN'S "Field Notes of Sutherlandshire," "The Gun," &c., J. JOHNSON, THE HON. GRANTLEY BERKELEY, "Greener on Gunnery," &c.

BREECH-LOADING CARTRIDGES,
AND CASES FOR SHOT GUNS AND RIFLES,
Primed with ELEY'S Double Waterproof Caps.

These Cases will be found very certain of ignition in all climates, and to stand the discharge without breaking or bursting, consequently they may be re-capped and loaded two or three times. If the chamber of the gun is correct in size and form, the empty cases will never stick in the breech, nor require any force to remove them—they should come out by a touch of the finger.

Cylindrical Wire Cartridges for BREECH-LOADING CASES.—These Cartridges will be found very effective in Breech-loading Guns, causing them to shoot with great additional strength and closeness.

Cartridges for ENFIELD and OTHER RIFLES.

Bullets for ENFIELD and other RIFLES, made by compression, with GREASED PATCH attached.

Cartridges made for Needle Rifles, for Rook and Rabbit Shooting—very simple and effective.

Patent Cartridges suitable for Adam's, Colt's, Deane-Harding, and Tranter's Revolvers.

Concaved Felt and Chemically prepared Cloth **Gun Wadding** warranted not to blow to pieces in the barrel, and every description of **SPORTING AMMUNITION** (Gunpowder excepted)

TO BE HAD OF ALL RESPECTABLE GUNMAKERS.

Eley Brothers, MANUFACTURERS, **London**

ELTON STORES: Darlington, Co. Durham.
Business: Stores.
Cartridges: Competition; Pest Control; Standard; Trapshooting.
Example: Competition. Ga 12; Tc crimson/black; De A flag flying on a staff; Kn Moisture resistant; Bl 8; Pr MCI; St Made in Gt. Britain. 12; Wd green/black. Special.Smokeless. 6; Case by Greenwood & Batley; Or British.

ELVEDEN ESTATE: Thetford, Norfolk.
Business: Private estate.
Cartridges: Elveden Estate.
Example: Elveden Estate. Ga 12; Tc light brown/black; Bl (Double head) 7 + 17; Pr MCI; St Cogswell & Harrison Ltd. № 12 Eley; Or British.

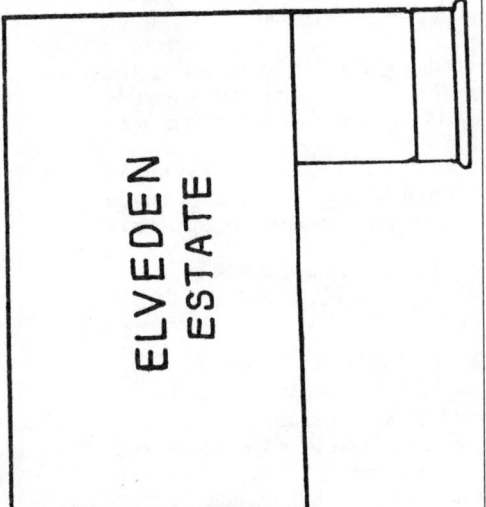

Case colour: Tan brown with black print.

DAVID EMSLIE: Elgin, Morayshire.
Cartridges: The Sniper.
Example: The Sniper. Ga 12; Tc vermilion red/black; Bl 8; Pr MCI; St ICI; Or British.

S. ENTWHISTLE: 151 Church St, Blackpool, Lancs.
Business: Gunsmith.
Cartridges:
Example: Unnamed. Ga 12; Tc red/black; De Cock pheasant standing with lowered tail, L; Kn Loaded by S. Entwhistle. Gunsmith. Phone, 2192; Bl 8; Pr MCI; St ICI; Wd red/black. S. Entwhistle.Blackpool. 6; Or British.

An earlier version of this cartridge with St KB, had the tube wording Kn Late, S. Troughton.

ERSKINE: Newton Stewart, Wigtowns.
Business: Gunmaker.
Cartridges: Information gleaned was that they made their own cases during the 1890s and then closed down around the time of the First World War.

BEN EVANS & CO: Swansea, Glam.
Cartridges: The Special.
Example: The Special. Ga 12; Tc blue/black; De A crest; Bl 8; Pr -CI; St NG; Or British.

C. A. EVANS: Burford, Oxon.
Business: Ironmonger.
Cartridges: The Cotswold.
Example: The Cotswold. Ga 16; Tc purple/silver; De A pheasant standing on the ground; Kn British loaded. Phone 35; Bl 8; Pr LCI; St SS; Wd red/black. Special · Smokeless · ; Or Foreign.

THOMAS J. EVANS: Welshpool, Mont (Powys).
Cartridges:
Example: Unnamed. Ga 12; Tc red/black; De Cock pheasant standing in grass with its tail in the horizontal position, L; Kn Specially loaded by Thomas J. Evans; Bl 8; Pr SCI; St T. J. Evans № 12 Welshpool; Or Not known.

WILLIAM EVANS: 63 Pall Mall, London. E.C.
Business: Gunmaker.
Cartridges: 20 Bore; 16 Bore; Ejector; Mark Over; Marlboro; Gastight; Pall Mall; Pegamoid; Sky High High Velocity.
Example: Pall Mall. Ga 12; Tc middle blue/black; Kn Telegrams, Shot Gun London; Bl 8; Pr LCI; St W. Evans № 12 London; Square turnover; Wd light blue/dark blue. W. Evans · London · 8; Or Not known.

J. W. EWEN: 45 The Green, Aberdeen.
Cartridges: The Ewen Special.
Example: The Ewen Special. Ga 12; Tc maroon/gold; De A hen pheasant, L; Bl 8; St SS; Or Foreign. No other details recorded.

EXPLOSIVES TRADES LIMITED: Witton, Birmingham. Also London.
Business: Ammunition and explosives manufacturers.
Cartridges: For details see Nobel Industries Limited.

Explosives Trades Ltd was the outcome of merging many firms, some small and some large, that were manufacturing explosives and their components. The end result was a brand new firm given the name Explosives Trades Limited. This came about in November 1918. Some of the cartridges they produced were given the small extra initials to the headstampings. These were E.T.L. For more details see, Nobel Industries Limited in this list. This is the name that the firm adopted two years later.

FAIRBURN: Guisborough, Yorks (Cleveland).
Cartridges:
Example: Unnamed or name not known. Ga 12; Tc red/black; De A pheasant; Pr MCI; St KB; Or British. No other details recorded.

FARMER: Leighton Buzzard, Beds.
Business: Gunmaker.
Cartridges: The ECEL.
Example: The ECEL. Ga 16; Tc maroon/gold; Bl 7. No other details recorded.

(*Below left*): The Kynoch's Patent Perfectly Gastight. Case colour: Light grey with Navy blue print.
Wad colour as illustrated: White card with black print.

FAWCETT: Kirkby Lonsdale, Westmorland (Cumbria).
Cartridges: The Lunesdale.
Example: The Lunesdale. Ga 12; Tc orange/black; Bl (Double head). No other details recorded.

FENWICK & SON: Stanhope, Co Durham.
Cartridges: Their name has been seen on an overshot wad; Wd white card/red.

FERRULES: The Arcade, Belfast, Northern Ireland.
Cartridges: No cartridge information available.

E. FLETCHER: Later E. FLETCHER & SON: Later still, FLETCHERS (SPORTS): 6 Westgate St, Gloucester, Glos.
Business: Gunmakers.
Cartridges: The Gloucester; The Pheasant Special.
Example one: Unnamed. Ga 20; Tc crimson/black; Kn Specially loaded by E. Fletcher. Gun maker. Gastight and metal lined; Bl 14; Pr MCI; St ELG; Square turnover; Wd white card/black. Shot size only, 5; Or British.
Example two: The Gloucester. Ga 12; Tc orange/black; Kn Special load. Fletchers (Sports) Ltd. Telephone, 22974; Bl 8; Pr MBI; St ICI; Wd brown/black (Cluster type) 6; Or British.

Another address seen on one of their cartridges was E. Fletcher & Son at 18 Westgate Street, Gloucester.

FOLLETT: Colyton & Seaton, Devon.
Cartridges: Their name has been seen on an overshot wad; Wd white card/red. This was loaded into a Tc brown. Eley case.

WILLIAM FORD: Eclipse Works, 15 St Mary's Row, Birmingham (Midlands).
Business: Gunmaker.
Cartridges: The Eclipse; The Fleet; Patent Ignition Tube.
Example: The Eclipse. Ga 12; Tc Gastight brick red/black; De Crest showing the sun in a partial eclipse; Kn Loaded by William Ford. Water resisting. Telegrams Eclipse. Telephone, Central 5210; Bl 16; Steel liner; Pr MCI; St ICI; Wd light blue/black. Chilled Shot. 3; Or British.

FORREST & SONS: Kelso-on-Tweed, Roxburgh.
Business: Gunmakers.
Cartridges: The Border Smokeless; The County; The Tweed.
Example: The Border Smokeless. Ga 12; Tc eau-de-nil/dark blue; A gun and rod forming a cross within an oval frame; Kn Loaded by Forrest & Sons, Kelso; Bl 8; Pr MCI; St ICI; Wd white card/red. Forrest & Sons · Kelso · 5; Or British.

RALPH FORTE: Oxford, Oxon.
Business: Ironmonger.
Cartridges: Information given to me was that cartridges had been seen bearing his name.

A. J. FOSTER: Sheffield House, 16 The Bull Ring, Kidderminster, Worcs.
Business: Gunsmith.
Cartridges: The Field Cartridge; The Quick Hit.
Example: The Field Cartridge. Ga 12, Tc greyish green/black; De The Ay Jay Effe Trade Mark; Loaded by A. J. Foster. Proprietor, S. Foster; Bl 8; Pr MCI; St ICI; Wd white card/red. Smokeless 5; Or British.

FOSTER LOTT & CO: The Ammunition Stores, Dorchester, Dorset.
Business: Ammunition sales.
Cartridges: Special Schultze Smokeless Cartridge.
Example: Special Schultze Smokeless Cartridge. Ga 12; Tc red/black; De A coat of arms. No other details recorded.

C. FOX: Canterbury, Kent.
Business: Gun and rifle maker.
Cartridges:
Example: Unnamed. Ga 12; Tc blue/black; Bl 8; St Smokeless 12 12 Gastight; Or Foreign. No other details recorded.

FOYS: Athlone, Westmeath, Republic of Ireland.
Cartridges: Smokeless Cartridge.
Example: Smokeless Cartridge. Ga 12; Tc orange/black; Kn Loaded by Irish Metal Industries: Bl 8; Pr MCI; St EK; Or Irish.

C. FRANCIS & SON: Peterborough, Northants (Cambs).
Business: Gunmakers.
Cartridges: The Demon.
Example: The Demon. Ga 12; Tc Gastight brick red/black; Kn Water resisting. Gastight. Metal lined. Tel 3152; Bl 16; Pr MCI; St ICI; Wd orange/black. Schultze 3; Or British.

FRANCIS & DEAN: St Mary's Hill, Stamford, Lincs.
Cartridges: Hy-Bird.
Example: Hy-Bird. Ga 12; Tc olive green/black; De Small standing cock pheasant, R; Kn Tel Stamford 279; Pr LBN; Steel head brass coated; St AZOT 12 66 Made in USSR. This is a Baikal imported cartridge. The 66 refers to 1966; Wd white card/olive green with the print forming a background. Shot size only, 5; Or Russian.

DANIEL FRASER & CO: Edinburgh & London.
Business: Gunmakers.
Cartridges: Pegamoid.
Example: Pegamoid. Ga 12; Tc brown/black; Kn Telegraphic address, Gunmakers Edinburgh. Telephone No 1012 Central. Pegamoid Brand paper; Steel lined; Bl 16; Pr MCI; St Daniel Fraser & Co Ltd. London & Edinburgh. No 12 Eley; Or British.

JOHN FRASER: Edinburgh, Midlothian.
Cartridges:
Example: Eley's Gastight Cartridge Case for Smokeless S. S. Powder. Ga 12; Tc grey/blue black; Bl 11; Pr SCI; St John Fraser No 12 Edinburgh. Eley; Or British.

NORMAN FRASER: Station Rd, Churchdown, Glos.
Business: Gunsmith.
Cartridges: Chosen.
Example: Chosen. Ga 12; Tc orange/black; Kn Name of cartridge is in longhand. Special load; Bl 8; Pr MCI; St ICI; Or British. Note: The name 'Chosen' was the abbreviated name for Churchdown.

"Chosen"
SPECIAL LOAD
Norman Fraser Churchdown Glos.

Case colour: Orange with black print.

FREENEY'S: High St, Galway, Republic of Ireland.
Cartridges: The Atom.

FRENCH & SON: Buckingham, Bucks.
Cartridges: Ejector.
Example: Ejector. Ga 12; One-piece brass case with a blue inner paper tube; Bl 57; Pr SCN; St NG; Wd white card/pink. French & Son · Buckingham · 6; Or Scottish; Name on wad only.

EDWARD FROST: Bridlington, Yorks (Humberside).
Cartridges:
Example: Unnamed. Ga 12; Tc Not recorded; Bl 16; St Smokeless 12 12 Gastight. No other details recorded.

S. C. FULLER: South St, Dorking, Surrey.
Business: Gunsmith.
Cartridges: The Long Shot.
Example: The Long Shot. Ga 12; Tc yellow/black; De A rabbit; St EL; Or British. No other details recorded.

This firm was established in 1897.

FUSSELL'S: London, Abergavenny, Port Talbot & Newport.
Business: Gunmakers.
Cartridges: The Club.
Example: The Club. Ga 12; Tc buttercup yellow/black; De A young cock pheasant walking on pine needles with tail slightly raised, L; Kn Non-rusting. Special Smokeless. British hand loaded. Addresses, 118–119 Cheapside, London E.C.2. 55 Cross St, Abergavenny. 81 Station Road, Port Talbot. 2 Dock Street, Newport, Mon. Foreign made case; Bl 8; Pr MCI; St SS; Wd light yellow/black. Special · Smokeless · 5; Or Foreign.

The Club. Case colour: Buttercup yellow with black print. Wad colour as illustrated: Primrose yellow with black print.

Believed to have been a Frank Dyke's load in an imported German case.

E. GALE & SONS: Joy St, Barnstaple. Also at 2–3 Mill St, Bideford, Devon.
Business: Gunmakers.
Cartridges: The County; The Field; The Flag; The X.L.
Example: The X.L. Ga 16; Tc middle blue/black; Bl 7; Pr MCI; St ICI; Or British.

GALLYON & SONS: Cambridge & King's Lynn. Later at, Peterborough.
Business: Gunmakers.
Cartridges: Camroid; Granta; Granton; Kilham; Lynton; Sandringham.
Example: Granta. Ga 12; Tc pale orange/black; Bl 8; Pr MCI; St EL; Wd orange/black. E. C. Powder. 5; Or British.

A. W. GAMAGE: Holborn, London. E.C.
Business: Retailers.
Cartridges: The A.W.G.; The Holborn; The Referee.
Example: Unnamed. Ga 20; Tc pale green/black; Bl 7; Pr MCI; St EL; Wd Colour not recorded. 'NE' Smokeless Powder. 6; Or British.

W. GARDEN: Aberdeen.
Cartridges: The Eclipse; Special Brown; The Granite City.
Example: The Eclipse. Ga 12; Tc signal red/black; Kn PH. 21050; Bl 8; Pr Medium Nickel Insert; St SS; Wd brown card/black. W. Garden's · Loading · 3; Or Foreign.

GARDENER: Chippenham, Wilts.
Cartridges: Warranted Gas-Tight Cartridge Case.
Example: Warranted Gas-Tight Cartridge Case. Ga 12; Tc olive green/black; Bl 11; Pr MCI; St Gardener № 12 Chippenham; Wd white card/red with the print forming a background for the white wording. Gardener Chippenham 5; Or Not known, believed British.

M. GARNETT: Crampton Court, Dublin, Republic of Ireland.
Cartridges:
Example: Unnamed. Ga 12; Tc burgundy/black; Kn Telephone 811; Bl Not recorded; Pr -CI; St RNC; Wd white card/red. Ruby · Smokeless · 6; Or U.S.A.

M. GARNETT & SON: 31 Parliament St, Dublin, Republic of Ireland.
Cartridges: The Kilquick; The Suredeath.
Example: The Kilquick. Ga 12; Tc dark green/black; Kn Damp proof. Smokeless Diamond powder velocity load. Telegrams Ammunition Dublin. A product of Irish Metal Industries Ltd. Manufactured under licence; Bl 16; Pr MCI; St EK; Wd red/black. Shot size only, 5; Or Irish.

FRANK GARRETT: Ilmington, Stratford-upon-Avon, Warwicks. Later at Evesham.
Business: Cartridge-loading expert.
Cartridges: Blue Flash Pigeon Cartridge; Crimson Flash; The D.B.H. (deadly but humane); Flash Junior ·410; Golden Flash; The Tempest.
Example: Crimson Flash. Ga 12; Tc pale yellow/crimson; De A lightning flash; Kn Loaded with the Patent Corona Wad; Bl 9; Pr MCI; St KB; Wd The light brown Patent Corona Wad. Raised moulded wording, Garretts Patent 4; Or British.

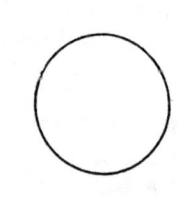

```
FRANK GARRETT'S
"FLASH JUNIOR"
EVESHAM
ENGLISH LOADED
```

(FOREIGN MADE CASE ·410)

Flash Junior. Case colour: Maroon paper tube with yellow print. Wad colour as illustrated: White card with no print.

ARTHUR GARRICK: Sunderland, Co Durham (Tyne & Wear).
Cartridges: The Sportsman.
Example: The Sportsman. Ga 12; Tc purple/silver; Bl 8; St SS. No other details recorded.

W. J. GEORGE: Dover, Kent.
Business: Gunmaker.
Cartridges:
Example: Unnamed. Ga 12; Tc brown tan/black; De Cock pheasant standing in grass with tail horizontal, L; Kn Guaranteed best loading; Bl 9; Pr MCI; St KB; Wd white card/red. W. J. George · Dover · 6; Or British.

GEVELOT: London.
Business: French ammunition manufacturers.
Cartridges: Solid Drawn Brass.
Example: Solid Drawn Brass. Ga 16; One-piece thin brass case; Bl 64; Pr LCN; St Gevelot 16 16 London; Or British/French. Note: The example is an unused case.

GEORGE GIBBS: 37 Baldwin St, Bristol. Also at 85 Savile Row, London.
Business: Gunmaker.
Cartridges: The Bristol; The County; The Farm Cartridge; The Field; The Gibbs.
Example: The Gibbs. Ga 12; Tc dark green/dark blue; Kn Keep dry; Bl 16; Pr MCI; St ICI; Wd white card/black. Gibbs · Bristol & London · 5½; Or British.

The Gibbs. Case colour: Dark or forest green with black or Navy blue print. Also seen with the wording 'Made in Great Britain' in small lettering. Wad colour as illustrated: White card with black print.

Case colour: Indian red with black print. Wad colour as illustrated: Signal red with black print.

GEORGE GIBBS: Reepham, Norfolk.
Business: Ironmonger.
Cartridges: Information that I received was that they once loaded cartridges.

GILL & CO: 5 High St, Oxford, Oxon.
Cartridges: The Dead Shot.
Example: The Dead Shot. Ga 12; Tc orange/black; De A running rabbit, R; Bl 8; St NG; Or Scottish.

JOHN H. GILL & SONS: Leeming Bar, London.
Cartridges: The Sproxton.
Example: The Sproxton. Ga 12; Tc orange/black; De Cock pheasant standing with a slightly raised straight tail, R; Kn John H. Gill & Sons (Leeming Bar) Ltd.; Bl 8; Pr ·MCI; Greenwood & Batley case; St MGB; Wd green/black. Special · Smokeless · 6; Or British.

C. GILLMAN & SONS: Black Jack St, Cirencester, Glos.
Business: Ironmonger.
Cartridges: Information given to me was that many years ago they sold a cartridge bearing their own name. They were supplied by Page-Wood of Bristol.

J. GILLMAN & SON: The Corner, Stafford St & Corporation St, Birmingham (Midlands).
Cartridges:
Example: Unnamed. Ga 12; Tc red/black; De Two opposed bantam-type birds standing on a convex ground line; Bl 8; Pr MCI; St ICI; Wd yellow/black. Smokeless · Diamond · 5; Or British.

GLIDDON & SONS: Williton, Somerset.
Business: Agricultural engineers.
Cartridges: The Exmoor.
I have been told that their name has been seen on a top wad with the colour white card/red. This was loaded into a plain Eley or Kynoch case.

G. E. GOLD: Castle Mill St, Bristol (Avon).
Cartridges: The Popular.
Example: The Popular. Ga 12; Tc green/black; Bl 8; St Gold Bristol. No other details recorded.

CHARLES GOLDEN: Bradford, Yorks.
Business: Gunmaker.
Cartridges:
Example: Unnamed. Ga 12; Tc brown/black; Bl 16; Pr MCI; St ELG; Wd orange/black. Schultze · * · 5; Or British.

WILLIAM GOLDEN: Huddersfield, Yorks.
Business: Gunmaker.
Cartridges: Ejector; Kynoch Waterproof Cartridge.
Example: Ejector. Ga 12; Two piece brass with a maroon inner paper tube; Bl 9 + 46; Pr MCI; St (Outer) Wm Golden № 12 Huddersfield. (Inner) Kynoch's Patent Grouse Ejector; Wd white card/red with the print forming a background for the white wording. Golden Huddersfield. (Inner) Schultze. 5; Or British.

G. E. GOLDING: Watton, Norfolk.
Business: Ironmonger.
Cartridges: The Wayland.
Example: The Wayland. Ga 12; Tc black/silver; Bl 16; St Smokeless 12 12 Gastight; Or Foreign.
  This cartridge took its name from Wayland's Wood, legendary site of the children's story 'The Babes in the Wood'.

JOHN R. GOW & SONS: Dundee, Angus.
Business: Gunmakers.
Cartridges: The Tayside.
Example: Unnamed. Ga 12; Tc orange/black; De Cock pheasant standing in grass, L; Kn Specially loaded by John R. Gow & Sons; Bl 8; Pr MCI; St ICI; Wd blue/black. Gow & Sons · Dundee · 5; Or British.

G. P. GRAHAM: Cockermouth, Cumberland (Cumbria).
Business: Gunmaker.
Cartridges: The Cumberland.
Example: The Cumberland. Ga 12; Tc Lichen green/black; De A rabbit bounding over grass, L; Kn Specially hand loaded by G. P. Graham; Bl 8; Pr MCI; St Graham Cockermouth № 12; Square turnover; Wd white card/nil. Blank load; Or Not known.

J. GRAHAM & CO: Union St, Inverness, Inverness-shire.
Business: Gunmakers.
Cartridges: Bon-Ton; Eley Ejector; The Highland; Pegamoid; The Primo; Special.
Example: The Highland. Ga 12; Tc burgundy/silver; Bl 16; St J. Graham & Co. Inverness. Eley; Or British. No other details recorded.

The Highland. Case colours: Dark or forest green with black or Navy blue print. Wad colour as illustrated: Deep blue with black print.

STEPHEN GRANT: 67A St James's St, London.
Business: Gunmaker.
Cartridges: The R.P.
Example: Unnamed. Ga 16; Tc spring green/black; Kn Patent Gas-tight Cartridge; Bl 11; Pr SCI; St (Reversed) Stephen Grant № 16 London; Wd white card/black. Special. Smokeless. 5; Or Not known.

Note: This case looks as though it may have been loaded in later years.

GRANT & LANG: 7 Bury St, St James's, London.
Business: Gunmakers.
Cartridges: The Briton; The Curzon; The Grantbury; The Instanter; Pegamoid; Rocketer; The Velogrant.
Example: The Briton. Ga 12; Tc pale green/black; Bl 8; Pr MCI; St KN; Or British.

Grant & Lang incorporated Charles Lancaster & Co and Watson Brothers. It ran from 1925 until 1960, when it became Atkin Grant & Lang.

D. GRAY & CO: 30 Union St, Inverness, Inverness-shire.
Business: Gunmakers.
Cartridges: Autokill; Ejector; Gastight; Pegamoid Waterproof.
Example: Autokill. Ga 12; Tc grey/black; Kn Loaded by Gray & Co. Telegrams, Sport, Inverness. Telephone 225 Inverness ; Bl 8; Pr MCI; St ICI; Wd green/black. Gray & Cº · Inverness · 6; Or British.

REG GRAY: Doncaster, South Yorks.
Business: Sports dealer.
Cartridges: The Don.
Example: The Don. Ga 12; Tc dark blue/black; De A standing cock pheasant with the tail high, R; Kn Anti-corrosive foreign cap and case. Loaded in Great Britain; Bl 8; Pr MCI; St Smokeless 12 12 Gastight; Or Foreign.

The Don. Case colour: Dark blue with black print.

EDWINSON GREEN & SONS:
Gloucester & Cheltenham Spa, Glos.
Cartridges: The Cotswold; Fur & Feather; Velox.
Example one: Velox. Ga 12; Tc pink/black; De All wording is within a large encirclement; Kn Smokeless Cartridge. British Make; Bl 8; Pr MCI; St KN; Wd orange/black. Greens Hand Loaded 5; Or British.
Example two: Fur & Feather. Ga 12; Tc middle blue/dark blue; De 'Fur & Feather' being written in longhand as a Registered Trade Mark; Kn British Make; Bl 16; Pr MCI; St ICI; Wd white card/black. Green's · Special Loading · 5; Or British.

W. W. GREENER: Birmingham, London & Hull.
Business: Gun and rifle maker.
Cartridges: Dead Shot; Ejector; Greener's Dwarf; Paragon; Police Gun E-K; Sporting Life.
Example one: Paragon. Ga 20; Tc orange/black; Bl 9; Pr MCI; St KN; Or British.
Example two: Sporting Life. Ga 24; Tc middle green/black; De A dark cock pheasant standing in grass, R. The wording 'Sporting Life' is printed on the pheasant's body; Kn London, Birmingham and Hull; Bl 9; Pr MCI; St W. W. Greener № 24 London & Birmingham; Or Not known.

W. W. Greener were in business between 1860 and 1966. They had their factory at St Mary's Square, Birmingham. They also had addresses at 68 Haymarket, London; 8 Avenue de l'Opéra, Paris; 38 Bolshaya Morskaya, St Petersburg, and at 176 Broadway, New York, U.S.A.

GREENFIELD: Storrington, Sussex.
Cartridges: Their name has been seen on an old cartridge stamping.

H. S. GREENFIELD: 4 Upper Bridge St, Canterbury, Kent.
Business: Gun and rifle maker.
Cartridges: The County Cartridge.
Example: The County Cartridge. Ga 12; Tc dark green/black; De A coat of arms being a shield with a crown above it and a scroll below. On the shield uppermost is a heraldry lion and below this are three standing birds facing left. The wording on the scroll is Ave Mater Angli; Kn Specially loaded by H. S. Greenfield; Bl 16; Pr MCI; St EN; Wd white card/black. Special Smokeless 5; Or British.

H. S. GREENFIELD & SON: 5 Dover St, Canterbury, Kent.
Business: Gunmakers and cartridge experts.
Cartridges: The County Cartridge.
Example: The County Cartridge. Ga 12; Tc pink or orange/black; De The coat of arms as above; Kn Expert loaded. Tel 2638. Estd, 1805; Bl 8; Pr MCI; St ICI; Or British.

GREENWOOD & BATLEY: Leeds & Farnham.
Business: Cartridge manufacturers.
Cartridges: A.E.C. Grey Squirrel; A.E.C. Pest Control; The Greenwood; The Skyrack; Standard Load; Trapshooting Cartridge; Trap Shooting Load.
Example: A.E.C. Grey Squirrel. Ga 12; Tc orange/black; Kn Made By Greenwood & Batley Ltd Leeds; Bl 8; Pr MCI; St MGB; Wd white card/black. Special · Smokeless · 6; Or British.

Cartridge cases were made in Ga 12, with either crimson or orange paper tubes. Cases and loads were supplied to other firms. On their own brands they used both rolled and star crimp closures.

GRENFELL & ACCLES: Perry Bar, Birmingham (Midlands).
Business: Cartridge manufacturers.
Cartridges:
Example: Unnamed. Ga 14; Tc white/nil; Bl Not recorded; Pr -CI; St Grenfell & Accles 14 Birmingham; Or British. No other details recorded. Note: A similar cartridge has been seen in Ga 12.

This firm was dissolved in 1896. See Accles Arms, Ammunition & Manufacturing Co. Ltd. at the start of this cartridge list.

GRIFFITHS: Manchester, Lancs (Gt. Manchester).
Cartridges: Eley's Gas-Tight Cartridge Case For E. C. Powder; Eley's Water-proof Cartridge Case From Pegamoid Brand Paper.
Example: Eley's, From Pegamoid Brand Paper. Ga 12; Tc brown/black; De The EBL shield with the wording Registered Trade Mark; Kn Patent Gas-tight. Water-proof; Bl 16; Pr MCI; St EL; Wd white card/red. Griffiths Manchester· 5½; Or British. Name on wad only.

S. J. GRIMES: Stamford, Lincs.
Business: Gun and ammunition depot.
Cartridges: The Champion; The Stamford Champion.
Example: The Stamford Champion. Ga 12; Tc deep red/black; De The brand name within a diamond-shaped frame; Kn Loaded in Great Britain. Phone 3146; Pr LC-; Bl 6; St SS; Wd red/black. Special · Smokeless · 4; Or Foreign.

THE GUN SHOP: Grantham, Lincs.
Cartridges: Eley's Ejector.
Name seen on the top wad.

GYE & MONCRIEFF: London.
Business: Gunmakers.
Cartridges:
Example: Unnamed. Ga 12; Tc blue/nil; Bl 10; St Messrs Gye & Moncrieff. No other details recorded.

C. HALL: Knaresborough, Yorks.
Cartridges: No cartridge information available.

FRANK HALL: Chesterfield, Derbyshire.
Cartridges: Hall-Right Special.
Example: Hall-Right Special. Ga 12; Tc buttercup yellow/black; Bl 7; Pr LCI; St SS; Wd white card/red. Frank Hall · Chesterfield · 6; Or Foreign.

JOHN HALL & SON: 79 Cannon St, London. E.C.4.
Business: Gunpowder manufacturers.
Cartridges: Believed that they may have loaded cartridges. Wad has been seen with the wording Hall's Loading. It could have belonged to a different Hall.

B. HALLIDAY & CO: 63 Cannon St, London. E.C.4.
Cartridges: The Express; High Velocity; Pheasant.
Example: The Express. Ga 12; Tc red/black; Kn Telephone, City 1085. Telegrams; 'Accurashot, Cannon, London.' (Near Queen Street); Bl 8; Pr MCI; St ICI; Wd yellow/black. Smokeless · Diamond · 6; Or British.

HAMMOND BROTHERS: Winchester, Hants.
Business: Gunmakers.
Cartridges: Pegamoid; The Reliance; Trusty Servant; Yellow Seal Mullerite (Hammond's name added to the tube printing).
Example: The Reliance. Ga 12; Tc light grey/black; Kn Loaded by Hammond Bros. Telephone 652; Bl 8; Pr MCI; St ICI; Wd yellow/black. Smokeless · Diamond · 5; Or British.

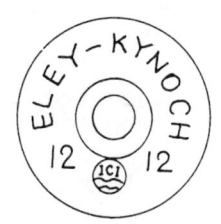

Trusty Servant. Case colours: Dark or forest green with black print. Also spring green with black print. Wad colour as illustrated: White card with red print.

W. T. HANCOCK: 308 High Holborn, London.
Cartridges:
Example: Unnamed. Ga 12; Tc dark green/black; Kn Kynoch's Patent Perfectly Gas-tight; Bl (Double head) 7 + 17; Pr SCI; St W. T. Hancock № 12 308 High Holborn; Wd white card/red with the printing forming a background for the white wording. Hancock Holborn 8; Or British.

HAND BROTHERS: High St, Odiham, Hants.
Business: Ironmongers.
Cartridges: The Pheasant.
Example: The Pheasant. Ga 12; Tc pale green/black; De A pheasant; Bl 16; St Believed to have been SS or Smokeless Gastight. No other details recorded.

F. G. HANDSCOMBE: Bishop's Stortford & Stansted, Hertfordshire.
Cartridges: Yellow Seal Mullerite. (Handscombe's name added to the tube printing).
Example: Yellow Seal Mullerite. Ga 12; Tc yellow/black; Bl 8; Pr LCI; St SS. No other details recorded.

HARDING BROTHERS: Hereford, Herefordshire.
Cartridges: The Rabbit Brand.
Example: The Rabbit Brand. Ga 12; Tc pink/black; De A small running rabbit; Kn Harding Bros Ltd; Bl 8; Pr MCI; St ICI; Wd brown card/black. Smokeless · Diamond · 5; Or British.

T. HARDING: Wiveliscombe, Somerset.
Business: Ironmonger.
Cartridges: Their name has been seen on a top wad with the colour white card/red. This was loaded into a plain case.

HARDY BROTHERS: Alnwick, Northumberland.
Business: Gunmakers.
Cartridges: Hardy's Northern; Hardy's Northern High Velocity; Hardy's Reliance.
Example: Hardy's Northern. Ga 12; Tc off-white/black; De Circular crest; Kn Smokeless Cartridge. Retail Depots, Edinburgh – 101 Princes St/Manchester – 12 & 14 Moult St/London – 61 Pall Mall, SW. By Appointment to H.M. the King; Bl 8; Pr MCI; St EN; Wd white card/black. Hardys · Nitro · 5½; Or British.

HARKOM & SON/SONS: Edinburgh, Midlothian.
Business: Gunmakers.
Cartridges: Ejector.
Example: Ejector. Ga 12; A two-piece brass case with a wine red inner paper tube; Bl 8 + 48; Pr Large SFM cap; St 12 12; Wd Colour not recorded. Name of firm on wad only; Or Foreign.

J. HARPER: Fosseway Garage, Cropwell Bishop, Notts.
Business: Motor garage.
Cartridges:
Example: Unnamed. Ga 12; Tc yellow/black; Kn The cartridge was a Mullerite Yellow Seal with additional tube printing; Bl 8; Pr LCI; St SS; Or Not known.

T. & W. HARRISON: Carlisle, Cumberland (Cumbria).
Cartridges: Name as seen on an over-shot wad. I have also seen the name of Harrison Bros. Carlisle on an old cartridge headstamping.

HARRISON & HUSSEY: 41 Albemarle St, London. W.
Business: Gunmakers.
Cartridges: The Albemarle; The Grafton; Stafford Deep Shell.
Example: The Grafton. Ga 12; Tc orange/black; Kn Telephone No 2300 Gerrard. Telegraphic Address, 'Weaponless, Piccy, London'; Bl 8; Pr MCI; St EN; Wd orange/black E.C. · Powder · 5½; Or British.

HARRODS: Brompton Rd, Knightsbridge, London. S.W.1.
Business: Large stores.
Cartridges: The Beaufort; The British Pioneer; The Kill-Sure; The Pioneer.
Example: The Kill-Sure. Ga 16; Tc red/black; De Trade Mark (Winged figure on the top of the world globe); Pr -CI; St EL; Or British. No other details recorded.

E. F. HART: Clare, Suffolk.
Cartridges:
Example: Unnamed. Ga 12; Tc orange/black; Bl 8; Pr MCI; St E. F. Hart. Clare. № 12 Eley; Wd Colour not recorded. E. F. Hart · Clare · 5; Or British.

F. W. HART: 39 Queen St, Scarborough, Yorks.
Cartridges: The Crackshot.
Example: The Crackshot. Ga 12; Tc orange/black; De Cock pheasant standing in grass with head high and his short tail lowered, L; Kn Telephone No 904; Bl 8; Pr MCI; St ICI; Wd brown card/black. Smokeless 5; Or British.

HARVEY GUNS: Great Yarmouth, Norfolk.
Cartridges: I have been told that a cartridge exists that has their name on it.

HARWOOD: The Square, Yarmouth, Isle of Wight.
Business: Ironmonger.
Cartridges: Information given to me was that cartridges were once loaded.

S. E. HAYWARD & CO: Tunbridge Wells & Crowborough.
Cartridges: The New Special Smokeless.
Example: The New Special Smokeless. Ga 12; Tc light purple/silver; De Cock pheasant walking on pine needles with head forward and tail slightly raised, L; Kn British hand loaded; Bl 8; Pr MCI; St SS; Wd yellow/black. Special · Smokeless · 5; Or Foreign.

G. HAWKE: St Austell, Cornwall.
Business: Ironmonger.
Cartridges:
Example: Unnamed. Ga 12; Tc yellow/black; Kn Mullerite Smokeless. G Hawke. St Austell; Bl 8; Pr LCI; St SS; Or Not known.

HAWKES & SONS: Taunton, Somerset.
Business: Ironmongers.
Cartridges: Hawk Brand.
Example: Hawk Brand. Ga 12; Tc's high-gloss greyish purple/silver, also seen orange/black; De Large perched hawk, L. The word Trade is to the left and the word Mark to the right and each within trimmings; Kn The name Hawk Brand is above the hawk in convex; Bl 8; Pr MCI; St EN; Wd orange/black. Schultze · ✱ · 8; Or British.

COLIN HAYGARTH: The Cottage Gunshop, Dunnet, Caithness.
Business: Gunsmith and cartridge loader.
Cartridges: The Economax.
Example: The Economax. Ga 12; Tc orange/black; De All the tube printing within a square; Kn Specially loaded. Tel, Barrock 802; Bl 8; Pr MCI; St MGB; Wd brown card/black. (Cluster type) 5; Or British.

WILLIAM HAYNES: 19 Duke St, Reading, Berks.
Business: Ironmonger.
Cartridges: I have in my collection a cartridge base. It is Bl 11; Pr MCI; St W. Haynes No 12 Reading. Haynes were in business during the 1920s.

W. E. HEAL: Bampton St, Tiverton, Devon.
Business: Ironmonger.
Cartridges: The Tivvy.
Example: The Tivvy. Ga 12; Tc Trafalgar dark blue/black; De Partridge with head high in vegetation, FR; Kn Special Smokeless Cartridge; Bl 8; Pr LCI; St Smokeless 12 12 Gastight; Wd red/black. Special · Smokeless · 5; Or Not known.

Hawk Brand. Case colour: Orange with black print. Wad colour as illustrated: Dark orange with black print.

The orange-cased Hawk Brand has also been seen on an Eley-Kynoch I.C.I. case. Also seen was a similar 12-bore cartridge with a high-gloss greyish purple paper tube having silver print. This also had an Eley Nobel headstamp.

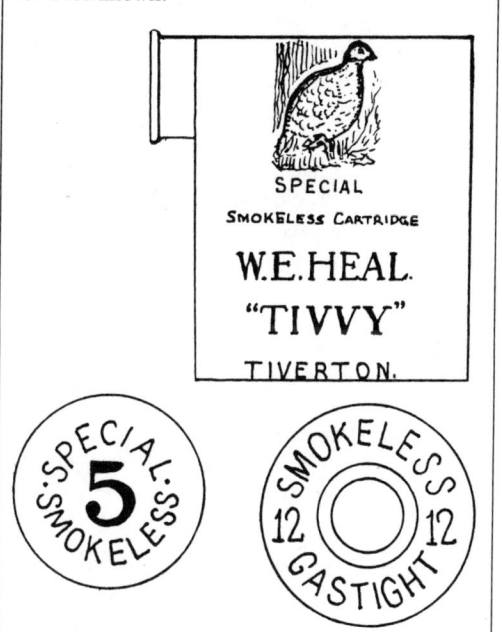

Tivvy Smokeless. Case colour: Dark or Trafalgar blue with black print. Wad colour as illustrated: Imperial red with black print.

T. HEATHMAN: Crediton, Devon.
Cartridges: Kynoch Witton Case.
Example: Kynoch Witton Case. Ga 12; Tc brown quality/nil; Bl 8; Pr SCI; St T. Heathman № 12 Crediton; Or British.

Note: This example is taken from my collection and is one of several that came to me from the U.S.A.

CHAS HELLIS & SONS: Edgware Rd, London. W.2.
Business: Cartridge experts.
Cartridges: 12 × 2 Inch; 12 × 2 Inch Deep Shell; The Burwood; The Championship; The Economist; The Edgware; The Falcon; The Merlin; Pegamoid; The Service; The Standard.
Example one: Unnamed. Ga 12; Tc purple/silver; Kn Specially loaded by C. Hellis & Sons. 119 Edgware Road; Bl 16; Pr LCN; St (Outer), Hellis & Sons-London 12. (Inner), Made in Belgium. Wd, yellow/black. Guaranteed · Accuracy · 4½; Or Belgium.
Example two: Unnamed. Ga 12; Cornflower blue/black or dark blue; De A standing woodcock, R; Kn Specially loaded by Chas Hellis & Sons Ltd. Cartridge Experts. 121-3 Edgware Road; Bl 16; Pr MCI; St ICI; Wd white card/black. Guaranteed · Accuracy · 5; Or British.

The firm of Charles E. Hellis was founded in 1884. His two sons Charles and Clifford came into the family business in 1902. After this date the words & Sons were added. Many different cartridges in numerous gauges were loaded, some of which did not carry brand names. Cartridges were also specially loaded to order. The Earl of Carnarvon at Highclere was one such customer. The over-shot wad used in these cartridges carried his personal crest. Their better-quality cartridges were usually fitted with large percussion caps. Cartridges were marketed from two addresses. These were 119 Edgware Road and later 121-3 Edgware Road. This firm closed its doors in 1956 and finally merged with Rosson's of Norwich and became Hellis-Rosson.

J. HELSON: 84 Fore St, Exeter, Devon.
Cartridges: The Demon; The Invincible.
Example: The Demon. Ga 12; Tc blue/black; De A running rabbit; St Helson, Exeter. Made in France; Or French. No other details recorded.

HELY: Dublin, Republic Of Ireland.
Cartridges: Hely's Rlymax.

HENDERSON: Dundee, Angus.
Cartridges:
Example: Unnamed. Ga 12; Tc light tan brown/black. Kn Kynoch's Perfectly Gas-tight Cartridge for E. C. Powder; Bl (Double head) 7 + 16; Pr SCI; St Henderson № 12 Dundee; Wd (May not be original) yellow/black. Special · Smokeless · 5; Or British.

HENRITE EXPLOSIVES: 97 Wilton Rd, London. S.W.
Business: Gunpowder and cartridge manufacturers.
Cartridges: Ejector; The Henrite.
Example: The Henrite. Ga 12; Tc buttercup yellow/black; Kn The main tube printing was the name Henrite in two directions as in a crossword so forming a cross on the letter R. Unusual is the minute wording printed in capital letters around the top on the roll of the turnover closure. This is Case Made In Bavaria; Bl 16; Pr LCI; St Henrite Explosives 12 12 London; Wd deep yellow/black. Henrite · * · 5; Or Bavarian. Note: Similar to above was also made in Ga 16. These cartridges were advertised in the 1907 catalogue of the Army & Navy C. S. Ltd, London.

ALEXANDER HENRY & CO: 12 Andrew St, Edinburgh, Midlothian.
Business: Gunmakers.
Cartridges: All their cartridges that I have seen did not have brand names.

W. G. HENTON & SONS: Lincoln, Lincs.
Cartridges: 'Special Loading'.
Example: 'Special Loading'. Ga, 12; Tc purple /silver; Bl 8; Pr MCI; St KB; Wd white card / Black. · Smokeless · 5; Or British.

T. HEPPLESTONE: Manchester, Lancs (Gt. Manchester).
Business: Gunmaker.
Cartridges:
Example: Unnamed. Ga 12; Tc brown quality/black; Bl 8; Pr SCI; St EL; Or British. The tube printing on this cartridge is towards the rolled closure.

HERCULES ARMS CO: 8 St Martin's, London. W.C.2.
Cartridges: The Farm Cartridge; The Hercules.
Example: The Hercules. Ga 12; Tc various colours, red, blue, light green, purple, yellow/black all on ribbed paper tubes; Kn Hand loaded by Hercules; Bl Pr Large Sinoxid copper cap (Rustless) with no insert cup; St Rustless 12 12 Waterproof; Wd (In purple case) mauve/black. Hercules · Waterproof · 6; Or Foreign.

HEWENS: Sheep St, Wellingborough, Northants.
Cartridges:
Example: Unnamed Ga 12; Tc brown/black; Kn Kynoch's Perfectly Gas-tight Schultze Cartridge; Bl 22; St Hewens No 12 Wellingboro; Or British. No other details recorded.

HICKLEY: Farnham, Surrey.
Business: Ironmonger.
Cartridges: Information given was that they once loaded their own cartridges.

FRED HICKS: 67 High St, Haverhill, Suffolk.
Business: Gunmaker and cycle agent.
Cartridges: Special Loading.

HARRY HIGGINS: 46–48 Teme St, Tenbury Wells, Worcs.
Cartridges: The Dead Shot; Harry Higgins Special.
Example: Harry Higgins Special. Ga 12; Tc pink/dark blue; De A dog at point, R; Kn Phone 110. Entirely British made. Better than ever; Bl 8; Pr MCI; St ICI; Or British.

(Below left): Harry Higgins Special. Case colours: Pink paper tube with either black or Navy blue print.

E. & G. HIGHAM: 4 Chapel St, Liverpool, Lancs (Merseyside).
Business: Gunmakers.
Cartridges: Gastight & Metal Lined.
Example: Gastight & Metal Lined. Ga 12; Tc green/black; Kn Established 1795; Bl 16; Pr -CI; St E & G Higham. Liverpool. № 12 Eley; Wd Colour not recorded. Higham · Liverpool · ; Or British.

G. G. HIGHAM: Oswestry, Salop.
Cartridges: Ejector; HIOS.
Example: Ejector. Ga 20; Two-piece brass case with an inner brown paper tube; Bl 9 + 48; Pr SCI; St Eley's No 20 Ejector. London.; Wd white card/red. Higham · Oswestry · ; Or British.

A. HILL & SON: 9 Market Place, Horncastle, Lincs.
Business: Gunmakers.
Cartridges: The Champion; The County Cartridge; The Ideal; The Reliable; Special Smokeless.
Example: The Reliable. Ga 12; Tc orange/black; Kn Telephone 3366; Bl 8; Pr MCI; St ICI; Or British.

A. Hill took over the business of G. H. Wilson at the same address in August 1902.

RUSSELL HILLSDON: Barnham, Chichester, Horsham & Worthing. Also at Birmingham.
Cartridges: The Combat; The Goodwood; The Revenge; The Sussex Champion; The Sussex Express.
Example one: The Sussex Champion. Ga 16; Tc middle blue/silver; Kn Hand loaded by Russell Hillsdon. Chichester & Barnham; Bl 7; Pr MCI; St EN; Wd powder green/black. (Outer), Russell Hillsdon. (Inner), Special Loading 6; Or British.
Example two: The Combat. Ga 12; Tc cream white/black; Kn Guaranteed accurately loaded of British components entirely. Chichester, Horsham & Worthing; Bl 8; Pr MCI; St ICI; Or British.

W. R. HINDE: Whitehaven, Cumberland (Cumbria).
Cartridges: Special Smokeless.

GEO HINTON & SONS: 5 Fore St, Taunton, Somerset.
Business: Gunmakers.
Cartridges: Special I.X.L.; The Standard; The Taunton.
Example: Special I.X.L. Ga 12; Tc eau-de-nil/dark blue; Kn Loaded by G. Hinton & Sons, Ltd; Bl 8; Pr MCI; St ICI; Wd yellow/black. G. Hinton · Taunton · 5; Or British.

The Standard was a Hinton's brand name at least since 1918. This business changed hands in 1947.

J. HOBSON: Leamington Spa, Warwickshire.
Business: Gunmaker.
Cartridges: The Challenge; The Dead Shot; Hobson's Choice; Hobson's Full Stop; Kynoch Grouse Ejector.
Example: Hobson's Full Stop. Ga 12; Tc off-white/black; De A rabbit curled up in the air with a peppered shot pattern; Bl 8; Pr MCI; St J. Hobson № 12 Leamington. Kynoch; Or British.

A. A. HODGSON: Louth, Lincs.
Cartridges: The Luda Cartridge.
Example: The Luda Cartridge. Ga 12; Tc dark red/black; De A shield-type crest; Kn Specially loaded by A. A. Hodgson; Bl 8; Pr MCI; St A. A. Hodgson № 12 Louth. Kynoch; Or British.

HODGSON: Bridlington, Yorks (Humberside).
Cartridges: This name has been seen on an overshot wad. White card/black with the shot size shown as IX.

HENRY HODGSON: Ipswich & Bury St Edmunds.
Business: Gunmaker.
Cartridges: The County De-Luxe; The Eclipse; The Express; The Special; Waterproof Cartridge Pegamoid.
Example one: Waterproof Cartridge Pegamoid. Ga 12; Tc brown/black; Kn Pegamoid Brand paper; Bl 16; Pr MCI; St ELG; Wd light orange/black. Schultze · ✽ · 6; Or British.
Example two: The Special. Ga 12; Tc greyish green/black; De A running rabbit, FL; Kn. ✽:✽. Tel Ipswich 2309. Tel Bury St Eds 559. Loaded with English powder; Bl 8; Pr MCI; St ICI; Wd orange/black. Hodgson · Bury St Edmunds· (Inner) Ipswich 5; Or British.

JESSE P. HODGSON: Louth, Lincs & Bridlington, Yorks (Humberside).
Cartridges: The name has been seen on an old Kynoch headstamping.

J. HODGSON: Lancaster, Lancs.
Cartridges: The Lancaster.
Example: The Lancaster. Ga 12; Tc light tan/black; Kn Special Smokeless; Bl 9; St SS. No other details recorded.

R. C. HODGSON: Ripon, Yorks.
Business: Gunmaker.
Cartridges: The Rapido.
Example: The Rapido. Ga 12; Tc orange/black; De A stroke of lightning; Bl 8; Pr MCI; St ICI; Wd yellow/black. Smokeless · Diamond · 5; Or British.

R. T. HODGSON: Station Bridge, Harrogate, Yorks.
Business: Ironmonger and gunsmith.
Cartridges: The Harrogate.
Example: The Harrogate. Ga 12; Tc crimson/black; De A coat of arms; Bl 16; Pr MCI; St R. T. Hodgson. № 12 Harrogate. Kynoch; Wd red/black. K. S. G. Powder; Or British.

W. HODGSON: Ripon, Yorks.
Business: Gunmaker.
Cartridges: The Rapido.
Example: Unnamed. Ga 12; Tc brown quality/black. Kn The tube printing W. Hodgson. Gunmaker. Ripon; Bl 8; Pr SBI; St W. Hodgson № 12 Ripon; Or Not known.

HOLDRON: Ashby-de-la-Zouch, Leics.
Business: Ironmonger.
Cartridges: The Rabbit.
Example: Unnamed. Ga 12; Tc brown quality/nil; Bl 8; St Holdron. Ashby. Eley; Or British. No other details recorded.

HOLLAND & HOLLAND: 98 New Bond St, London. W.1.
Business: Gunmakers.
Cartridges: The Badminton; The Badminton High Velocity; The Badminton High Velocity Large Cap; The Dominion; Ejector; High Velocity; Nitro Paradox; Pegamoid; Recoilite; The Royal; The Twelve-Two;
Example one: Unnamed. Ga 2; 4½ inch length case (115 mm) with reinforced paper at the base end of tube; Kn The tube printing is diagonal and upside down towards the top. Eleys' Gastight Cartridge Case; Tc dark green quality/black; Bl 18 including the rim 3·5; Pr SCI; St Holland & Holland · № 2 · ; Wd white card/shot size only inked with pen, black ink. BB; Or British.
Example two: The Badminton Cartridge. Ga 16; Tc brown tan/black; Bl 9; Pr MCI; St Holland & Holland № 16 Eley; Or British.
Example three: The Royal. Ga 12; Case length unloaded, 2½ inch (65 mm); Paper-tubed cartridge with an extended inner metal liner and crimped over for the closure. The paper length from the top of Bl is 1½ inches (39 mm). The extended metal liner is ⅜ inch (10 mm): Tc gastight brick red/black; Kn (Br Patent No 494264); Bl 16; Pr MCI; St ICI; Or British.
Example four: Recoilite. Ga 16; Tc darkish blue/nil; Bl 16; Pr MCI; St ICI; Wd white card/red. Recoilite · Regd · 6; Or British.

Note: This name was on the over-shot wad only. Although the name of Holland & Holland was not printed on the cartridge, I was told that their name was printed on the box. During the Second World War many firms sold plain cased cartridges for civil use and tube printing was not always carried out. Often the firm's name and the cartridge brand name was printed on a box wrapper and the box was often plain.

The Royal. Case colour: Eley's Gastight brick red with black printing. This case has an internal alloy metal tube which extends 10 mm above the outer paper tube. The cartridge is closed by a special crimping of the extended metal tube.

HOLLIS: address not known.
Cartridges: Hollis Special Cartridge.
Example: Hollis Special Cartridge. Ga 20; Tc salmon pink/black; De Royal warrant coat of arms; Kn This case shows no address or town; Bl 7; Pr MCI; St EL; Or British.

T. J. HOOKE: Later, T. J. Hooke & Son. 38–39 Pavement, Coppergate, York, Yorks.
Business: Gunmakers.
Cartridges: Eclipse Cartridge.
Example one: Unnamed. Ga 12; Tc pale orange/black; De Cock pheasant with tail horizontal, L; Kn T. J. Hooke. Gunmaker. 38–39 Pavement; Bl 8; Pr MCI; St T. J. Hooke № 12 York; Or Not known.
Example two: Eclipse Cartridge. Ga 12; Tc purple/silver; Kn Name of the cartridge is convexed and in capitals. Gunmakers. Coppergate, York. English loaded; Bl 8; Pr LCI; St (Outer) Special 12 12 Smokeless. (Inner) Foreign Made Case; Wd white card/red. Special · Smokeless · 5; Or Foreign.

HOOTON & JONES: 60 Dale St, Liverpool, Lancs (Merseyside).
Cartridges: Hooton & Jones's Special.
Example: Hooton & Jones's Special. Ga 12; Tc and Bl Not recorded; St Hooton & Jones № 12 Liverpool. Eley; Wd pinkish buff/black. Schultze · * · (Shot size only) 1×; Or British.

J. J. HOPKINS: 2 – 4 Lake St, Leighton Buzzard, Beds.
Business: Ironmonger.
Cartridges: The Golden Pheasant.
Example: The Golden Pheasant. Ga 12; Tc maroon/gold; De A golden hen pheasant with raised tail walking on pine needles, L; Kn Special smokeless. British loaded. Metal lined; Bl 8; Pr MCI; St Smokeless 12 12 Gastight; Or Foreign.

The Golden Pheasant. Case colour: Maroon with gold print. Wad colour as illustrated: Light yellow with black print.

THOMAS HORSLEY & SON: Micklegate, York, Yorks.
Business: Gunmakers.
Cartridges: Gastight; Horsleys' Smokeless Rabbit Cartridge.
Example: Horsleys' Smokeless Rabbit Cartridge. Ga 12; Tc dark red/black; De A running rabbit with no ground drawn in, L; Kn Practical gunmaker. Case made in Belgium; Bl 9; Pr LCN; St In large square-type lettering. Horsley 12 12 York. Wd orange card/nil; Or Belgium.

W. HORTON: 199 Buchanan St, Glasgow, Lanarkshire.
Business: Gunmaker.
Cartridges: The Horton Cartridge; Ejector; Extra; Weatherproof.
Example: The Horton Cartridge. Ga 12; Tc light brick/black; Bl 16; St Horton Glasgow. No other details recorded.

HOWARD BROTHERS: 240 St Ann's Rd, Tottenham, London.
Cartridges: The only information I have is that they had their name on cartridges.

HULL CARTRIDGE CO: 58 De Grey St, Hull, Yorks (Humberside).
Business: Cartridge loading and wholesale.
Cartridges: The Standard; The Three Crowns.
Example: The Three Crowns. Ga 12; Tc crimson/black; De The Three Crowns shield; Kn Cartridge case made by Eley-Kynoch; Bl 16; Pr MCI; St ICI; Wd dark green/black. Special · Smokeless · 7; Or British. This firm occupies premises that once were Turners Carbides. The Three Crowns, which is a Trade Mark, has been loaded in several different cases in Ga's 12, 16 and 20. The Hull Cartridge Co still load their own cartridges and also many brands for private firms. They have also been known to load on specially printed cases for private individuals when a sufficiently large order has been placed. Many of their loadings have been in the British Eley and the Italian Fiocchi cases.

HUNTER & MADDIL: 58 Royal Ave, Belfast, Northern Ireland.
Business: Gunmakers.
Cartridges: I have no details of their cartridges. I believe that they were in business between 1890 and 1920.

HUNTER & SON: Belfast, Northern Ireland.
Business: Gunmakers.
Cartridges: De Luxe; The Eclipse; Express Cartridge; The Favourite; The Ideal; The Invictus; The Long Shot; The Reliable; The Royal Cartridge; The Universal.
Example: The Long Shot. Ga 12; Tc dark blue/black; Kn Metal lined; Bl 8; Pr LCI; St Smokeless 12 12 Gastight; Or Foreign.

HUNTER & VAUGHAN: Broad St, Bristol (Avon).
Cartridges: Special Smokeless.
Example: Special Smokeless. Ga 12; pale yellow/black; Bl 8; Pr MCI; St KB; Or British.
  This firm incorporated Septimus Chamber & Co.

H. J. HUSSEY: 81 New Bond St, St James's. Also at 88 Jermyn St, London.
Business: Gunmaker.
Cartridges: Eley's Ejector; The Times.
Example: The Times. Ga 12; Tc brown tan/black; Bl 16; Pr MCI; St EL; Wd pale orange/black. Hussey Ltd. 88 Jermyn Street. S.W. 5½; Or British.

HUTCHINGS: Aberystwyth, Cardigan (Dyfed).
Business: Gunmaker.
Cartridges: Special Smokeless Cartridge.
Example: Special Smokeless Cartridge. Ga 12; Tc salmon pink/black; Kn (Additional wording) Kynoch's Patent Perfectly Gastight; Bl (Double head) totals = 21; Pr MCI; St KB; Or British.

HUTCHINSON: Kendal, Westmorland (Cumbria).
Business: Gunmaker.
Cartridges: Nobel's Sporting Ballistite Special Cartridge.
Example: Nobel's Sporting Ballistite Special Cartridge. Ga 12; Tc yellow/black; De A standing partridge, L; Kn Nobel's Sporting Ballistite Special Cartridge. Hutchinson. Gunmaker. Kendal; Bl 10; Pr MCI; St Nobel's Ballistite. Nº 12 Eley; Wd Colour not recorded. Hutchinson · Kendal · 4; Or British.

HUTCHINSON ROE & CO: Stone St, Cranbrook, Kent.
Cartridges: The Knockout.
Example: The Knockout. Ga 12; Tc purple/silver; De A running rabbit; Bl 8; St SS; Or Foreign. No other details recorded.

IMPERIAL CHEMICAL INDUSTRIES (I.C.I.) METALS DIVISION, ELEY-KYNOCH: Witton, Birmingham (W. Midlands).
Business: Ammunition manufacturers and marketers.
Cartridges: 2 Inch Deep Shell; 4 Gauge; 8 Gauge; 10 Gauge; 20 Gauge; 20th Century; 20th Century Deep Shell; Acme; Almac (Brass ·410); Alphamax; Alphamax High Velocity; Alphamax Neoflak; Blank; Bonax; British Smokeless Cartridge; British Smokeless Case; Coronation Cartridge; Deep Shell; Ejector; Ejector Thin Brass; Eley Half Brass; Empire; Extra Long (3 inch ·410); Fourlong (2½ inch ·410); Fourten (2 inch ·410); Gastight Cartridge; Gastight Case; Gastight Quality Case; G.P.; Grand Prix Cartridge; Grand Prix High Velocity Cartridge; Grand Prix Case; Grand Prix Quality Case; Hymax; Impax; Maximum High Velocity; Maximum Neoflak; Mettax; Nitrone; Noneka; Parvo; Pegamoid Cartridge; Pegamoid Case; Primax; Rocket; Saluting Blank; Saluting Blanks; Scarebird; Smokeless Cartridge; Sporting Ballistite; Trapshooting; Two-Inch; Universal (Kyblack); Velocity; Westminster; Wildfowling; Winchester Cannon; Yeoman; Zenith.
Example one: Grand Prix Cartridge. Ga 12; Tc Nearly always orange/black. Also I have seen blue, pink, purple, red, white and yellow/black; De The EBL shield Trade Mark; Kn Eley loaded. Smokeless Diamond; Bl 8; Pr MCI; St ICI; Wd brown card/black. Eley-Kynoch Loaded 5; Or British. Note: Similar cartridge loaded in Ga 16; Tc dark blue/black.
Example two: Coronation Cartridge. Ga 12; Tc Trafalgar blue/silver; De The coronation crown; Kn This cartridge has no EBL Shield Trade Mark but it does have a small five-pointed star; Bl 8; Pr MCI; St ICI; Wd deep red/black. Eley Loaded. Smokeless Diamond. 7; Or British.
Example three: Rocket. Ga 12; Tc grey/dark blue; De The EBL shield with the wording, Registered Trade Mark. Also there is a shot trail with an exploding star and the wording Shows The Flight Of The Shot; Kn Patented throughout the World. Or Patent applied for; Bl 8; Pr MCI; Rim has a knurled edge; St ICI; Wd blue card/black. Eley Loaded. Smokeless Diamond. 8; Or, British.
  Note: Similar cartridge seen in Ga 12 with Tc orange/black. Also a Government load in Ga 12 with Tc cream white/red. These had the Government's broad arrow on the tube print. Example

three was also loaded in Ga's 16 and 20.

Example four: 2" G.P. Ga 12; Tc red/black; De EBL shield Trade Mark; Kn For 2 inch chambered guns. Made in Great Britain. Around and just below the rolled closure printed four times is 2 inch; Case length, 2 inches (51 mm); Bl 8; Pr MCI; St ICI; Wd yellow/black. Smokeless · Diamond · 5; Or British.

This very large company took over from Nobel Industries Limited in 1926. From the beginning of their control, all the manufacturing of shotgun ammunition was undertaken from the old Kynoch Lion Works at Witton, Birmingham. Although we here are only concerned with sporting shotgun cartridges, this company also made rifle, pistol and many other kinds of military ammunition. They also made many items for the explosives trade for military and civil uses.

In the shotgun line, they made pinfire cartridges with plain white paper tubes with no printing in the Ga's 12 and 16. These were often used in alarm guns or loaded with black powder and shot. Some other cartridges made by them were as follows: A.E.C. Grey Squirrel; A.E.C. Pest Control; A.E.C. Rook; Agricultural Departments Reserve; Forestry Pest Control; W.A.E.C. Grey Squirrel. Some cartridges were specially loaded for the British Government during World War Two. The L.D.V. (Local Defence Volunteers), later renamed the Home Guard, were given cartridges in Ga 12 in variously coloured paper tubes with the broad arrow printed in black. These were often very heavy loads and also Lethal Ball. Special Trapshooting was a cartridge produced with the broad arrow added to its tube printing. It was made in Ga 12 with the Tc being white or orange/black. This cartridge was used for initial training of airgunners by the R.A.F. It is possible that the other services may have used it as well. Also made for military use were signal cartridges, engine start cartridges and cartridges to scare birds off aerodromes. Some of these that I have just mentioned have also found their place in civvy street. Marshall Tractors required special Ga 12-loaded Tc green/black cartridges to start them.

As can be seen by this cartridge list, I.C.I. also provided cases and loaded very many cartridge brands to the order of other firms in the British Isles. They also exported this service as well. All this firm's cartridge cases, except those that had special stampings to order, and larger in gauge than ·410, had the I.C.I. Trade Mark on their headstampings. The only exceptions were some for Northern Ireland and the Australian and New Zealand markets through their own overseas factories. The headstamping Eley-Kynoch I.C.I. was only discontinued when I.C.I. themselves were taken over by Imperial Metal Industries (I.M.I.) about 1962. I do not intend to concern myself to any degree with this later Division within this cartridge list.

At a glance of the listed cartridges, one can see how I.C.I. honoured the names of many of the firms that had been absorbed into this large concern. The brand names of some of these old cartridges still provided good sales for the I.C.I. company.

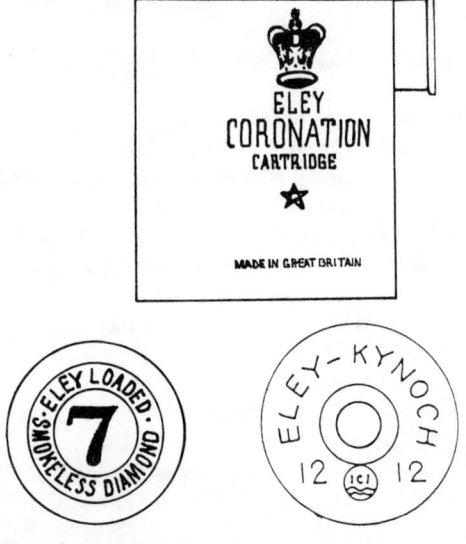

Eley Coronation Cartridge. Case colour: Navy blue with silver print. The cartridge tube is covered with a high-gloss varnish that gives it a greyish tint. Wad colour as illustrated: Imperial red with black print.

W. H. ICKE: Smithford St, Coventry, Warwickshire (Midlands).
Business: General furnishing and ironmonger.
Cartridges:
Example: Unnamed. Ga 12; Tc brown/black; De A crest. This being a form of shield with an elephant and a lion over the top of it; Kn Kynoch's Perfectly Gastight; Bl (Double head) 6 + 19; Pr -CI St W. H. Icke № 12 Coventry; Or British.

CHARLES INGRAM: 10 Waterloo St, Glasgow, Lanarkshire.
Business: Gunmaker.
Cartridges: The Ingram.
Example: The Ingram. Ga 12; Tc eau-de-nil/dark blue; Bl 8; Pr MCI; St ICI; Or British.

IRISH METAL INDUSTRIES: Dublin & Galway, Republic of Ireland.
Business: Ammunition manufacturers.
Cartridges: Alphamax; Maximum; Primax.
Example: Primax. Ga 12; Tc dark green/black; Kn Kynoch Primax Cartridge. Smokeless. A product of Irish Metal Industries Ltd. Manufactured under licence; Bl 16; Pr MCI; St EK; Wd brown card/black. Eley-Kynoch * IMI * 5; Or Irish.

Although I only know the three brand names that I have listed above I feel sure that this is only the tip of the iceberg. This firm was started by I.C.I. Metals Div. Eley-Kynoch back in the 1930s.

DAVID IRONS & SONS: Forfar, Angus.
Business: Ironmongers.
Cartridges:
Example: Unnamed. Ga 12; Tc orange/black; De Cock pheasant standing; Bl 8; Pr MCI; St ICI; Or British.

JACKSON: Gainsborough, Lincs.
Cartridges: An old gentleman once told me that he had used their cartridges.

H. G. JACKSON: Bungay & Halesworth, Suffolk.
Business: Ironmonger and gunsmith.
Cartridges:
Example: Unnamed. Ga 12; Tc purple/silver; De A large crown; Kn Loaded by H. G. Jackson. Ironmonger and gunsmith. Bungay & Halesworth; Bl 8; Pr -CI; St Jackson № 12 Bungay & Halesworth; Wd Colour not recorded. (Shot size only) 5; Or Not known.

S. JACKSON: Church Gate, Nottingham, Notts.
Business: Gunmaker.
Cartridges: The Nottingham.
Example: The Nottingham. Ga 12; Tc buff/black; De The Nottingham coat of arms; Bl 10; St Jackson № 12 Nottingham. No other details recorded.

JAMES & CO: Great Western Mills, Hungerford, Berks.
Business: Millers and game food manufacturers.
Cartridges: The Kennett.
Example: The Kennett. Ga 12; Tc red/black; Kn Smokeless; Bl 7; Pr MCI; St SS; Wd white card/red. Special · Smokeless · 5; Or Not known.

M. JAMES & SONS: Newcastle Emlyn, Carmarthen (Dyfed).
Business: Ironmongers.
Cartridges: The Gwalia.
Example: The Gwalia. Ga 12; Tc red/black; De A rabbit bounding along flat out over ground, L; Kn Special smokeless; British hand loaded; Bl 8; Pr LCI; St Jas R. Watson & Co 12 12 London; Wd dark blue/black. Cooppal Smokeless 4; Or Foreign.

W. H. JANE: Bodmin, Cornwall.
Business: Ironmongers and gun dealers.
The example below has been seen and it is believed that it was theirs. The tube printing is worn away in places.
Example: The Bodmin. Ga 12; Tc maroon/silver; De A running rabbit, L; Kn Special smokeless. Phone 55?; Bl 8; St SS; Wd yellow/black. Special · Smokeless · ; Or Foreign.

A. R. & H. V. JEFFERY: 100 Old Town St, Plymouth, Devon. Also at Yeovil.
Cartridges: The Empire.
Example: The Empire. Ga 12; Tc orange/black; De A standing bird looking much like a snipe, R; Kn Specially loaded by A. R. & H. V. Jeffery Ltd. Telephone No 60187. Telegraphic address, Guns Plymouth; Bl 8; Pr MCI; St ICI; Or British.

C. JEFFERY & SONS: Dorchester, Dorset.
Business: Gunmakers.
Cartridges: The Ejector; The Empire; The Rabbit; The Royal Game; The Twenty.
Example: The Twenty. Ga 20; Tc buff/black; Kn Telephone No 32 (Night & Day); Bl 7; Pr MCI; St ICI; Or British.

The Royal Game. Case colour: Dark or forest green with black print. Wad colour as illustrated: White card with red print.

S. R. JEFFERY & SON: Guildford & Salisbury.
Business: Gunmakers.
Cartridges: 12-Bore Pinfire; Eley's Gastight Cartridge Case; ·410 (Short); ·410 (Long); The Champion; The Club; Smokeless Powder; Special Smokeless.
Example one: Unnamed. Pinfire. Ga 12; Tc green quality/black; Kn The diagonal tube printing is Eley's Gastight Cartridge Case; Bl 11; A very small rim; St Jeffery No 12 Guildford; Or British.
Example two: The Champion. Ga 12; Tc orange/black; Kn Smokeless. Loaded by S. R. Jeffery & Son Gunmakers Guildford; Bl 8; Pr MCI; St KN; Wd light brown card/black. Schultze Powder 6; Or British.

W. JEFFERY & SON: 3 Russell St, Plymouth, Devon.
Business: Gunmakers.
Cartridges: The Eddystone; The Pegamoid; The Rabbit; The Sky High.
Example: The Sky High. Ga 12; Tc black/silver; De A walking pheasant, L; Kn British hand loaded. Phone 1477; Bl 8; Pr MCI; St SS; Wd white card/red. Smokeless 5; Or Foreign. Note: A similar cartridge Tc black/silver was loaded in Ga 16.

W. J. JEFFERY & CO: St James's, London.
Business: Gunmakers.
Cartridges: The Champion; The Club Smokeless; The Ejector; High Velocity; The Jeffery Cartridge; Jeffery's XXX; The Sharpshooter.
Example one: The Champion. Ga 12; Tc maroon/silver; De Two circles one within the other forming a crest. This holds the majority of the wording; Kn 13 King St. St James' St, London; Bl 16; Pr MCI; St W. J. Jeffery & Co № 12 London. E; Wd buff/black. Schultze · * · 5½; Or British.
Example two: Club Smokeless. Ga 20; Tc red/black; De Trade Mark. This consisting of the letter J in a circle; Kn 60 Queen Victoria St London. E.C. Loaded in London; Bl 7; Pr LCI; St W. J. Jeffery & Cº London. № 20, Made in Belgium. Wd red/black. Smokeless 6; Or Belgium.
Example three: Club Smokeless. Ga 12; Tc yellow/black; De Trade Mark. This consisting of the letter J in an oval; Kn 26 Bury St, St James's, London; Bl 8; Pr LCI; St W. J. Jeffery & Cº. Ltd 12 12 London · Made in Bavaria; Wd scarlet/black. Smokeless 3; Or Bavarian.
Example four: Sharpshooter. Ga 12; Tc red/black; De A round crest with the wording within it For One And All; Kn 9 Golden Square, Regent Street, London. W.1; Bl 8; Pr MCI; St EN; Wd yellow/black. Precision · Loading · 7; Or British.
Note: Four examples have been shown and each one has a different address.

A. J. JEWSON: 1 Westgate, Halifax, Yorks.
Business: Gunmaker.
Cartridges: The Champion; The Crown; The Westgate.
Example: The Crown. Ga 12; Tc gastight brick red/black; De Two crowns with a shield between them; Kn Specially loaded by Jewsons. Practical gunmaker. Phone 4146; Steel liner; Bl 16; Pr MCI; St ICI; Or British.

G. JOBSON: Milford, Surrey.
Cartridges: The Milford.
Example: The Milford. Ga 12; Tc maroon/gold; De Cock pheasant walking with tail high; Kn Special smokeless. British loaded. Phone, Godalming 266; Bl 8; Pr MCI; St SS; Wd yellow/black. Special · Smokeless · 6; Or Foreign.

T. JOHNSON & SON: Swaffham, Norfolk.
Business: Gunmakers.
Cartridges: Ejector; Johnson's Celebrated Ne Plus Ultra; The Reliable.
Example: Johnson's Celebrated Ne Plus Ultra. Ga 12; Tc purple/silver; De A flying pheasant; Bl 16; Pr MCI; St NGN; Or Scottish.

JOHNSON & WRIGHT: Northampton, Northants.
Cartridges: The County.
Example: The County. Ga 12; Tc dark orange/black; De A coat of arms consisting of two lions, one each side of a castle tower; Kn Specially loaded for Johnson & Wright Limited; Bl 16; Pr MCI; St (Outer), Johnson & Wright. Ltd. No 12 (Inner), Kynoch Northampton; Wd red/black. · K.S.G. · Powder 5; Or British.

WILLIAM JOWETT: 3 Kingsbury, Aylesbury, Bucks.
Business: Ironmonger.
Cartridges: The Kingsway.

FREDERICK JOYCE & CO: 57 Upper Thames St. Also at 7 Suffolk Lane, London. E.C.
Business: Cartridge case manufacturers.
Cartridges: Bailey's Gastight Cartridge; The Bonnaud; Ejector; Solid Drawn Case; F. J. Gastight; Ideal Smokeless; Improved Gas-tight for Amberite Smokeless; Improved Gas-tight for Schultze; Improved Gas-tight for S.S.; Improved Gas-Tight for Walsrode; Joyce's D. B. Cartridge; Special Nitro; Special Smokeless; The Waltham.
Example: Unnamed. Pinfire. Ga 16; Tc olive green/blue black; Kn Joyce's Improved Gas-tight Cartridge; Bl 9; Extra small rim; St Joyce & Co No 16 London; Or British.
    F. Joyce & Co were cartridge case manufacturers and made very strong cases. They made pin- and central-fire cases in various sizes. Cases were sold to private firms for loading and also marketed overseas. Many Joyce cases and cartridges had no tube printing. During 1907 the Nobel Explosives Co took complete control of F. Joyce & Co. They then used this firm for the supply of their cartridge cases. A close examination of Joyce and Nobel's cases will prove the point.

JULIAN: Basingstoke, Hants.
Business: Gunsmith, cycle and hardware store.
Cartridges:
Example: Unnamed. Ga 12; Tc brownish orange/black; Kn The inverted diagonal tube printing is Eley's Gas-tight Cartridge Case for EC Gunpowder; Bl 10; Pr SCI; St Julian · Basingstoke · No 12 Eley; Or British.

W. KAVANAGH & SON: Dublin, Republic of Ireland.
Business: Gunmakers.
Cartridges: The Ideal; The Mirus; The Schultze Cartridge.
Example: The Ideal; Ga 20; Tc blue/black; Kn Selected smokeless. Loaded by W. Kavanagh & Son. Dublin; Bl 15; Pr SCI; St Kavanagh & Son 20 20 Dublin; Wd white card/black. (Cluster type) 5; Or Not known.

L. KEEGAN: 3 Inns Quay, Dublin, Republic of Ireland.
Cartridges: The Faugh-A-Ballagh.

ALFRED KENT & SON: Wantage, Berks (Oxon).
Business: Furnishing and ironmongers.
Cartridges: Kynoch's Perfectly Gas-tight; The Wantage.
Example one: Perfectly Gas-tight. Ga 12; Tc brownish tan/black; Kynoch's Patent Perfectly Gas-tight charged with E.C. Gunpowder. Specially loaded by Kent & Son, Wantage; Bl (Double head) 6 + 17; Pr SCI; St Kent & Son No 12 Wantage; Or British.
Example two: Unnamed. Ga 12; Tc pale greyish green/black, or eau-de-nil/dark blue; De Cock pheasant standing on a grass mound with tail out straight, L; Kn Kent & Son, Ironmongers, Wantage. Telephone No 41; Bl 8; Pr MCI; St ICI; Wd white card/red. Smokeless 5; Or British.

KENYON & TROTT: Cattle Market, Ipswich, Suffolk.
Business: Now believed to be a plating firm.
Cartridges:
Example: Unnamed. Ga 12; Tc middle green/nil; Bl 4; Pr SCI; St No 12; Wd white card/red with the print forming a background for the white wording. Kenyon & Trott · Ipswich ·; Or Not known.

KENT & SON,
Ironmongers, ✠ Wantage,

Have a Large and Assorted Stock of

SPORTING REQUISITES.

AGENTS FOR

Kynoch's WORLD RENOWNED **Cartridges.**

Also Cases & Wadds of every description.
Extractors, Cleaning Rods, &c.
Cartridge Bags, Belts, &c.

Cartridges loaded with Black or any Kind of Smokeless Powders with any size Shot.

GUNS & RIFLES LENT ON HIRE.

Fishing Tackle, Footballs, Lawn Tennis Racquets & Balls, Golf, Croquet, Quoits, &c.

Special Double Barrel Farmer's Central Fire Breech Loading Gun. Warranted. Price 55/-.

Co-operative Stores beaten by our Special Cheap 12 bore C. F. Cartridges, our own Loading, with Best Treble Strong Black Powder and any size Shot, 6/9 per 100. Cash with order. Every Cartridge warranted.

> **Memorandum**
>
> From **ALFRED KENT & SON**, WHOLESALE & RETAIL Furnishing & General Ironmonger, LAMP & OIL MERCHANT, WANTAGE.
>
> To _____ 18__
>
> SIR,
>
> We are supplying a good quality 12 gauge green Central Fire Cartridge, loaded with treble strong Black Powder and any size Hard Shot, at 6/9 (six shillings and ninepence) per 100, Cash with order. We also load Cartridges with any special Powder, such as E.C., Schultze, &c., in either Eley, Joyce, or Kynoch Cases, at proportionate prices.
>
> Your early orders will much oblige,
>
> Yours obediently,
>
> KENT & SON.
>
> *Guns and Rook Rifles for Sale and Hire, and every Requisite for Sporting.*

CHARLES KERR: Stranraer, Dumfries & Galloway.
Business: Gunmaker.
Cartridges: The Royal.
Example: The Royal. Ga 16; Tc red/black; De The royal coat of arms; Kn Estd 1810; Bl 8; Pr MCI; St KB; Or British.

H. E. KERRIDGE EXORS: Great Yarmouth, Norfolk.
Business: Ironmonger and gunsmith.
Cartridges: The Champion; The East Anglian.
Example: The Champion. Ga 12, Tc blue/gold; De A walking cock pheasant, L; Bl 8; Pr LCI; St Kerridge 12 12 Yarmouth; Or Not Known.

KINGDON: Basingstoke, Hants.
Business: Hardware stores.
Cartridges:
Example: Unnamed. Ga 12; Tc dark green/black; De The EBL shield with the outlined lettering and rope-type edging and the wording Registered Trade Mark; Kn The tube printing is Eley's Gas-tight Cartridge Case. Made in Great Britain; Bl 10; Pr MCI; St Kingdon. Basingstoke. Nº 12 Eley; Wd orange/nil; Or British.

JAMES KIRK: 36 Union Buildings, Ayr, Ayrshire.
Business: Gunmaker.
Cartridges: The Champion; The High Velocity; The Land of Burns; The Marksman; The Retriever; Special Pegamoid; Special.
Example: The Land of Burns. Ga 12; Tc greyish green/black, or eau-de-nil/dark blue; Kn Specially loaded by James Kirk. Smokeless. Phone 3390; Bl 8; Pr MCI; St ICI; Wd white card/black. J. Kirk. Ayr · Smokeless · 8; Or British.

THOMAS KIRKER: Belfast, Northern Ireland.
Business: Gunmaker.
Cartridges: I have been given their name but do not have any details of their cartridges.

H. KIRMAN: Scunthorpe, Lincs (Humberside).
Business: Ironmonger.
Cartridges: Farmers Special.
Example: Farmers Special. Ga 12; Tc yellow/black; De A running rabbit, L; Bl 8; Pr MCI; St KB; Or British.

KITHER: Sevenoaks, Kent.
Cartridges:
Example: Unnamed. Ga 12; Tc light brown/black; Kn Eley's Gastight Cartridge Case for Schultze Sporting Powder; Bl 10; Pr MCI; St ELG; Wd white card/red. Kither · Sevenoaks · ; Or British. Note: Name on wad only.

J. N. KNIGHT: Wells, Somerset.
Business: Ironmonger.
Cartridges:
Example: Unnamed or name not known. Ga 12; Tc brown/black; St KB; Or British. No other details recorded.

P. KNIGHT: Clinton St, Nottingham, Notts.
Business: Gunmaker.
Cartridges: The Castle; The Invincible; The Thurland.
Example: The Invincible. Ga 12; Tc dark green/black; De Picture of a knight with a shield portrayed standing upright when the cartridge is standing on its base. With capital letters placed below each other are the words Knights on the left and Invincible on the right; Bl 16; Pr MCI; St EN; Square-type roll closure; Wd white card/red with the print forming a background for the white wording. Knight Nottingham 6; Or British.

GEORGE KYNOCH & CO: The Lion Works, Witton, Birmingham (W. Midlands).
Business: Ammunition, candle and cycle manufacturers.
Cartridges: Blue Quality; Brown Quality; Green Quality; Maroon Quality; Gastight Waterproof; Warranted Gastight; Perfectly Gastight; $^{5}/_{16}''$ Brass; $^{5}/_{8}''$ Brass; The Bonax; The C.B.; Deep Shell; The Kyblack; The Kynoid; The Nitro Ball; The Nitrone; The Opex; The Paradox Bullet Cartridge; Patent Perfect Metallic; Patent 2090 Grouse Ejector; The Primax; The Sallinoid Ball Cartridge; The Swift; The Tellax; The Triumph; The Witton.
Example one: Blue Quality. Pinfire. Ga 20; Tc dark blue/nil; Bl 12; Very small rim; St KB; Wd brown card/black. (Shot size only) 6; Or British.
Example two: Kynoch's Patent Perfectly Gastight. Ga 4; Tc dark green/black; Kn Diagonally printed was the above name; Case length 4 inches (103 mm); Bl (Double head) 16 + 25, this includes the rim 3; Pr SCI; St KB; Wd white card/nil; Or British.
Example three: The Triumph. Ga 12; Tc dark bluish green/black; De A crown; Kn Perfectly Gastight Case for Walsrode; Bl 26 (1 inch); Pr MCI; St KB; Or British.
Note: Similar cartridges with Tc orange/black were for Mullerite and Tc cream yellow/black for Ballistite powder.

George Kynoch started the cartridge business in 1862 at Witton in a wooden hut. From this modest beginning Kynoch emerged to become a limited company in 1884 manufacturing ammunition on a scale similar to Eley Bros.

Kynoch's made cartridges in many sizes, ranging from Saloon (Garden Gun) in sizes 1, 2 and 3 in short and long through the normal gauges to puntgun size. Some of the curiosities which have come to light are an all-brass "Perfect" in 18-bore, and several brass pinfires in a variety of bores. Their very early pinfire cases were Tc brown/nil, they then introduced their green and blue quality-tubed cartridges.

Apart from marketing their own loaded cartridges, they also did a good trade in ready-primed cases in this country and abroad. Firms would often have the tubes printed for them, and expensive cartridges would have names and towns stamped on the heads. Various Gastight-quality cases were made and printed to suit the numerous powders on the market, in both black and nitro. The priming cap for each type was made to match the appropriate powder. The powders were Amerite, Ballistite, Cannonite, E.C., K.S.S. (Kynoch's Smokeless Sporting), Nonpareil, Schultze, Shotgun Rifleite, S.S., Walsrode and T.S. In the early days the Brown Quality was considered a cheap case and was later introduced as the Witton. Some, but not all, had the name Witton on the headstampings. Green was often referred to as 'Extra Quality' and blue as 'Best Quality'. Last of the Kynoch unprinted casts were the maroon ones.

The Opex, an ejector type, had external brass that protruded above the rolled turnover of the paper tube; its brand name was also on the headstamping. The famous Patent 2090 was produced during 1886 and it included the doubling of the brass head. The Swift was listed in 1911 and was advertised along with other brands on

Following pages:
1 *A famous Yellow Wizard stands out amongst these cases*
2 *A few famous British boxes*
3 *A varied selection of cases, including an all-brass case*

W. JEFFERY & SON
BRITISH LOADED
R. BLANTON
GUNMAKER
RINGWOOD
PHONE 223
YELLOW
WIZARD
SMOKELESS
A. F. PUNTER
Proprietor: J. M. EMBERTON
46 WOTE STREET
BASINGSTOKE
KENT & SON
Ironmongers
WANTAGE
Telephone No. 41
W. E. HEAL
THE
"CUMBERLAND"
SMOKELESS CARTRIDGE
Specially Hand Loaded
VELOX
CARTRIDGE
SMOKELESS
GLOUCESTER & CHELTENHAM
BRITISH HAND LOADED
M. JAMES & SONS.
IRONMONGERS
NEWCASTLE EMLYN
FOREIGN MADE
THE NEW
LIMITED
COMPANY
Specially loaded by
G. R. WEBSTER
30a Wide Bargate
BOSTON
Phone: 3124
GUNMAKERS
"PERFECTION"
SMOKELESS
CARTRIDGE
4 MARKET HEAD
BY APPOINTMENT
The "NORFOLK"
(WATER-RESISTING)
SEE OUR
"TWELVE-TWENTY"

MIDLAND CARTRIDGE CO.
SPECIAL
RABBIT CLEARANCE SOCIETY CARTRIDGE
ECONOMY
25
REMINGTON UMC
12GA.
BRITISH LOADED
WETPROOF
SMOKELESS CARTRIDGES
THE REMINGTON ARMS UNION METALLIC CARTRIDGE COMPANY LTD.
BRIMSDOWN, ENGLAND.
NEW YORK
LONDON
25
25
"Kelor"
"SPECIALLY TESTED"
CARTRIDGES
COGSWELL & HARRISON
LIMITED.
UNLINED. SHALLOW BRASS
LONDON
SMOKELESS
CARTRIDGE
MADE IN ENGLAND
TRENT GUNS & CARTRIDGES, LTD.
SHOT TOWER
GRIMSBY
THOS. TURNER & SONS LTD.
GUNMAKERS
86 Northbrook St., NEWBURY, Berks
SMOKELESS CARTRIDGES
"FILLBAG"
THOS. TURNER & SONS LTD
NEWBURY & BASINGSTOKE
25
ELEY
ACME
CARTRIDGES
LOADED WITH
POWDER
E.C.
POWDER
12
GAUGE
5
SHOT
COGSWELL & HARRISON
LTD.
Established 1770
25
25
TESTED CARTRIDGES
"MARKOR"
SMOKELESS POWDER
BRITISH MADE THROUGHOUT
ELEY
Velocity
CARTRIDGE
See that
STEEL LINING
TRENT
TRENT
BRITAIN'S
CARTRIDG
MADE IN ENGLAND
CHEAPEST AND BES
PNEUMATIC
CARTRIDGES
THE MOST EFFICIENT YET EVOLVED
BONAX
CARTRIDGES
Loaded with
EC
Made at the
KYNOCH FACTORIES
of
IMPERIAL CHEMICAL INDUSTRIES LTD
LONDON, ENGLAND
NO BLOWBACK
NO SMOKE
25
REDUCED HEAT
QUICK IGNIT
SPECIAL SMOKELESS
HAND
LOADE
GASTIGHT CARTRIDGES
LOADED BY THE
MIDLAND GUN CO
BIRMINGHAM
ENGLAN

ELEY SMOKELESS
Cartridges
25
12 GAUGE
LOADED WITH SCHULTZE POWDER
25
NOBEL'S
"CLYDE"
SMOKELESS
CARTRIDGES
GAUGE
12
NOBEL INDUSTRIES LTD
LONDON
25
12GA.
REMINGTON UMC
NITRO CLUB
SHOTGUN CARTRIDGES
Smokeless Powder
LOADED BY
THE REMINGTON ARMS UNION METALLIC CARTRIDGE COMPANY LTD.
AT BRIMSDOWN, ENGLAND.
NEW YORK
LONDON
GREENER
"Sporting Life"
CARTRIDGES
BORE
POWDER
SHOT.
SPARTAN
DEEP SHELL
SPARTAN
DEEP SHELL
BEST SMOKELESS
MADE IN ENGLAND
"PHEASANT"
DEEP SHELL
CARTRIDGES
B. HALLIDAY & CO. LTD.
GUNMAKERS
63 CANNON STREET
LONDON, E.C.4
ELEY
WATER PROOF
CARTRIDGES
LOADED WITH
SMOKELESS ◆ DIAMOND
POWDER
PECAMOID
25
ELEY
12 Gauge
UNLINED NITRO
CARTRIDGES
LOADED WITH
CURTIS'S & HARVEY'S
SMOKELESS DIAMOND
25 ELEY 12 GAUGE
WESTMINSTER
SMOKELESS
CARTRIDGES
Inter fulmina securus
LOADED WITH
SCHULTZE POWDER
25
ELEY
GAUGE
SPECIAL
TRAPSHOOTING
CARTRIDGES
SPECIALLY
LOADED FOR
CLAY TARGET
SHOOTING
25
12
GAUGE
ELEY
ELEY
"MAXIMUM"
LONG RANGE
MAXIMUM
CRIMP CLOSURE
ELEY LOADED
WITH A NOBEL
SMOKELESS POWDER
WATER-RESISTING
BY APPOINTMENT
RIFLE MAKERS
TO HER MAJESTY THE QUEEN
JOHN RIGBY & CO.
GUN AND RIFLE MAKERS
CARTRIDGES
32 KING STREET
ST. JAMES'S
LONDON, S.W.1
Telephone
TRAfalgar 5212

GAS-TIGHT
CARTRIDGE CASE
Made in Great Britain
SMOKELESS
36
N E
GRAIN
POWDER
NITRO
LONG RANGE
EXPRESS
MADE IN U.S.A.
ELEY
GASTIGHT
WATER RESISTING
CASE
Remington
Smokeless
WILLIAMS & POWELL,
GUNMAKERS.
GAS-TIGHT
CARTRIDGE CASE
KYNOCH
"PERFECTLY PATENT GASTIGHT."
SMOKELESS
CARTRIDGE
METAL LINED
SELECTED MATERIAL
BY
GEORGE BATE
(GUNMAKERS) LTD.
132 STEELHOUSE LANE
BIRMINGHAM
ELEY
10 GAUGE
CARTRIDGE
ELEY LOADED
WITH A NOBEL
NOBEL'S
SPORTING BALLISTITE"
Special Cartridge

the Kynoch Display Mirrors.

Most of the Kynoch brand names lasted until November 1918, when the firm merged with others to become Explosives Trades Ltd. (See under Eley). The manufacture of shotgun cartridges continued from the Lion Works, and in 1920 the firm changed its name to Nobel Industries Ltd. The later Kynoch brand names such as Bonax, Deep Shell, Kyblack, Nitrone and Primax were kept long after the cessation of Kynoch.

An extract from an old Kynoch book, '*Shooting Notes and Comments*' published in 1910, states: "Opex (1st quality) Metal extends beyond turnover. Kynoid (2nd quality) Waterproof (Paper) case. Primax (3rd quality). Bonax (4th quality). Tellax (5th quality) Loaded with just under 1 oz of shot: Introduced to meet cheap competition of foreign rubbish that has been sold in this country. The shooter who buys a cheaper round than this does so at the risk of losing eye or fingers".

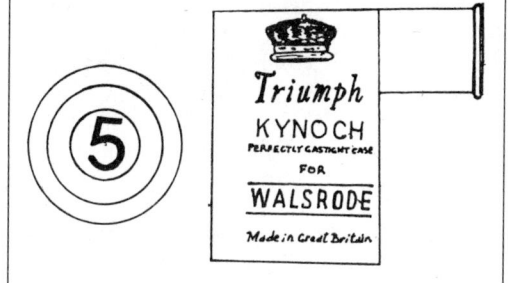

(*Below left*): Case colours: Walsrode, peacock blue. Ballistite, straw yellow. Mullerite, orange. Print is black.

LACE: Market Place, Wigan, Lancs (Gt. Manchester).
Cartridges: Lace's Smokeless Cartridge.
Example: Lace's Smokeless Cartridge. Ga 12; Tc light pink/black; De The EBL shield Registered Trade Mark with rope-type edging; De Lace's Smokeless Cartridge is written diagonally similar to that on the Eley Brothers' Smokeless Cartridge; Kn Eley. Lace's Smokeless Cartridge. Market Place. Wigan; Bl 8; Pr MCI; St EL; Wd pink/black. Eley's · Smokeless · 5; Or British.

ARTHUR LACEY: Bridge St, Stratford-upon-Avon, Warwickshire.
Business: Ironmongers.
Cartridges: The Welcome Smokeless.
Example: The Welcome Smokeless. Ga 12; Tc red/black; De A standing cock pheasant with tail horizontal, L; Kn Specially loaded by Arthur Lacey; Bl 9; Pr MCI; St Lacey · Stratford on Avon · № 12 Kynoch; Or British.

CHARLES LANCASTER: London. S.W.
Business: Gunmaker and cartridge developer.
Cartridges: Ejector; Generally Useful; The Leicester; The Norfolk; Patent Gastight Cartridge; The Twelve Twenty.
Example one: The Leicester. Ga 12; Tc olive green/black; Kn 11 Panton Street, London. S.W. Gunmaker to H.M. King George V; Case length 2¼ inches (56 mm); Bl 16; Pr MCI; St Eley N.I. № 12 London; Wd white card/black. Charles Lancaster 6; Or British.
Example two: The Norfolk. Ga 12; Tc light brown/black; De The royal coat of arms by appointment. Water resisting. See our Twelve-Twenty gun. 99 Mount Street, London. W.1. Tel Oval bore London. Telephone, Grosvenor 1394; Bl 16; Pr MCI; St ICI; Wd white card/black. Charles Lancaster 7; Or British.

Charles Lancaster was one of the pioneers in cartridge case development. He patented his own case to prevent blowbacks. This had the cap hidden behind the thin brass base. His London addresses were 11 Panton Street, circa 1904–1925. Then a move was made to 99 Mount Street, circa 1925–1932. The next move was to 151 New Bond Street. This firm, then as Charles Lancaster & Company Limited, was later incorporated into Grant & Lang along with Watson Brothers.

LANE BROTHERS: Faringdon, Berks (Oxon).
Business: Ironmonger and agricultural merchants.
Cartridges: The Eclipse.
Example: The Eclipse. Ga 12; Tc crimson/black; Kn Ironmongers; Bl 25 (Single head); Pr MCI; St KB; Wd red/black. Wording not recorded; Or British.

FRANK LANE & CO (FARINGDON): Faringdon, Berks (Oxon).
Business: Ironmongers and agricultural merchants.
Cartridges:
Example: Unnamed. Ga 12; Tc pink/black or dark blue; De Cock pheasant standing with tail horizontal, L; Kn Specially loaded for Frank Lane & Co (Faringdon) Ltd; Bl 8; Pr MCI; St ICI; Wd orange/black E. C. Powder 8; Or British.

JOSEPH LANG & SON: 10 Pall Mall, London. Also at 102 New Bond St, London.
Business: Gunmakers.
Cartridges: Ejector; Lang's Special; The Ventracta.
Example: The Ventracta. Ga 12; Tc dull red/black; Kn 10 Pall Mall; Bl 11; St Lang. 10 Pall Mall. No other details recorded.

JOHN LANGDON: 20 St Mary St, Truro, Cornwall.
Business: Gunsmith.
Cartridges: The Langdon.
Example: The Langdon. Ga 12; Tc light blue/silver; De A standing cock pheasant, L; Kn Phone 307. British hand loaded; Bl 8; Pr -C-; St SS; Or Foreign.

LANGLEY: Hitchin, Herts. & Luton, Beds.
Business: Gunmaker.
Cartridges: Blue Roc; Kynoch Patent Grouse Ejector; The Prize Winner.
Example: Blue Roc. Ga 12; Tc brownish orange/black; De A large dove standing upright, L. with the words Reg Trade Mark; Kn Kynoch's Patent Perfectly Gas-tight; Bl (Double head) 7 + 17; Pr MCI; St Langley № 12 Luton; Or British.

Langley's cartridges have been seen in various different cases as for some reason he kept seeking fresh sources of supply. He later joined forces with Aubrey Lewis and for a time the firm became Langley & Lewis.

LANGLEY & LEWIS: Park Square, Luton, Beds. & Maldon, Essex.
Business: Gunmakers.
Cartridges: Blue Roc; British Smokeless; Prize Winner.
Example: Blue Roc. Ga 12; Tc light brown/black; De A large standing dove, L; Kn Langley's Blue Roc. Metal lined. Telephone 619 Luton; Bl 16; Pr LBN; St Langley & Lewis № 12 Luton. (Inner) Case Made In France; Or French.

TOM LAW: Castle Douglas, Kirkudbrightshire.
Cartridges:
Example: Unnamed. Ga 12; Tc dark green/black; De A standing capercaillie, L; Kn Loaded by Tom Law; Bl 16; Pr MCI; St ICI; Wd orange/black. T. Law · Castle Douglas · 5; Or British.

LAWN & ALDER: address unknown.
Cartridges: L. & A.
Example: L. & A. Ga 12; Tc shocking pink/black; Kn Waterproof. Specially loaded for Lawn & Alder; Bl 12; Pr MCI; St Remington UMC № 12 Nitroclub; Wd white card/black. (Shot size only) 8; Or U.S.A.

I have been told by several people that they think this is a British firm, but I have no history of it.

R. LEACH: Oldham, Lancs (Gt. Manchester).
Cartridges: Eley Gastight.
Example: Eley Gastight. Ga 12; Tc gastight brick red/black; Kn Metal lined; Bl 16; Pr MCI; St ICI; Wd Colour not recorded. R. Leach · Oldham · ; Or British.

LEE: Bishop's Stortford, Herts.
Cartridges:
Example: Unnamed or name not known. Ga 12; Tc brown quality/black; Bl 8. No other details recorded.

LEECH & SONS: Chelmsford, Essex.
Business: Gunmakers.
Cartridges: The Chelmsford; The Club Smokeless; Essex County; Kynoch Patent No.2090 Grouse Ejector; Leech's Special Load; The X.L.
Example: The Club Smokeless. Ga 12; Tc greyish green/black; De A black club as on a playing card; Kn Loaded by Leech & Sons; Bl 8; Pr MCI; St ICI; Wd red/black. Leech & Sons · Chelmsford · 6; Or British.

W. R. LEESON: Ashford, Kent. Also, London.
Cartridges: The Invicta.
Example: The Invicta. Ga 20; Tc dark buff/blue; The Kent prancing horse with the word "Invicta" below on a scroll; Kn Ashford & London; Bl 16; Pr Not recorded; St W. R. Leeson Ltd No 20 Ashford; Wd white card/black. W. R. Leeson Ltd · Ashford · 6; Or Not known.

LEIGH & JACKSON: Witney, Oxon.
Business: Ironmongers.
Cartridges: An old cartridge base has been found. Ga 12; Bl 8; St Leigh & Jackson. Witney. № 12 Eley; Or British.

L. LePERSONNE & CO: 7 Old Bailey, London. E.C.4.
Business: Wholesalers.
Cartridges: I do not know if they ever put their name on a cartridge, but they did market Lepco and F. N. cartridges in the British Isles. There was an alloy-covered cartridge called The Metalode with a Belgian appearance which may have been marketed by them.

AUBREY LEWIS: 19 Church St, Luton, Beds.
Business: Gunmaker.
Cartridges: Blue Roc; The Chelt; Fourten; Eley G.P.; High Velocity; The Severn; The Special.
Example: Eley G.P. Ga 12; Tc orange/black; De The EBL shield Trade Mark; Kn (Extra tube printing) Specially loaded by Aubrey Lewis. Gunmaker. Luton; Square turnover; Bl 8; Pr MCI; St ICI; Wd orange/black. Aubrey Lewis. Luton. 6; Or British.

Aubrey Lewis was in partnership with Langley as Langley & Lewis. Aubrey later took over full control and continued in business until he finally closed his doors in 1969.

G. E. LEWIS & SONS: Lower Loveday St, Birmingham (W. Midlands).
Business: Gunmakers.
Cartridges: The Express; The Keepers Cartridge; Pegamoid Case; The Premier.

J. D. LEWIS: Narberth, Pembrokeshire (Dyfed).
Business: Ironmongers.
Cartridges: This shop was known to have loaded their own cartridges.

TED LEWIS: Basingstoke, Hants.
Business: Gunsmith.
Cartridges:
Example: Unnamed. Ga 12; Tc black/silver; De A running rabbit; Bl 8; Pr LCI; St SS; Or Not known.

Note: The example was taken from an unused ready capped case, believed to have come from the Trent Gun & Cartridge Works of Grimsby. A similar cartridge has been seen loaded in a yellow tubed case.

The Lewis Cartridge. Case colour: Black with silver print.

LIDDELL & SONS: Haltwhistle, Northumberland.
Cartridges: I have been given an overshot wad. This is red/black. Liddell & Sons · Haltwhistle · 5.

LIGHTWOOD: Bournemouth, Hants (Dorset).
Cartridges:
Example: Unnamed. Ga 12; Tc blue quality/nil; Bl 9; Pr SCI; St Lightwood № 12 Bournemouth; Loaded with an Eley wire cartridge; Believed to be an Eley Brothers case and British.

LIGHTWOOD & SON: Price St, Birmingham, Warwickshire (W. Midlands).
Cartridges: The Ecel.

F. W. LIGHTWOOD: Brigg, Grimsby & Market Rasen, Lincs & Humberside.
Business: Gunmakers.
Cartridges: The Four Best; Woodcraft.
Example: Woodcraft. Ga 12; Tc brown/black; De A sitting dog with a bird in its mouth; Bl 16; Pr MCI; St KB; Or British.

S. J. LIMMEX & CO: Wood St, Swindon, Wilts.
Business: Gunsmiths and ironmongers.
Cartridges: No details recorded as a cartridge I was to be shown was stolen.

LINCOLN JEFFRIES: 121 Steelhouse Lane, Birmingham, Warwickshire (W. Midlands).
Business: Gunmakers.
Cartridges:
Example: Unnamed. Ga 12; Tc orange tan/black; Kn Loaded by Lincoln Jeffries. Gold medal, London; Bl 8; Pr MCI; St EL; Or British.
 Note: This example was an unused capped case.

G. LINES: Stevenage, Herts.
Cartridges:
Example: Unnamed. Ga 12; Tc black/silver; De A running rabbit. No other details recorded.

J. H. LININGTON: Newport, Isle of Wight.
Business: Ironmonger.
Cartridges: Extra Special; Pinfire.
Example: Extra Special. Ga 12; Tc deep mauve/black; De A crest, this being a shield with a sailing ship within; Bl 16; Pr MCI; St Linington. Newport.I.W. № 12 Eley; Wd mauve card/nil; Or British.
 Note: This example is taken from a window cartridge. It is a dummy display with an oval-shaped celluloid window built into its paper tube showing the loaded contents. One or two of these display rounds have come to light but I have never seen a fired or live cartridge of the type.

LINSLEY BROTHERS: Lands Lane and 97 Albion St, Leeds. Also at Bradford, Yorks.
Business: Gunmakers.
Cartridges: Ejector; High Velocity; Nomis; Standard; The Swift.
Example: Unnamed. Ga 20; Tc brown/black; De The Leeds coat of arms with the three owls; Kn Lands Lane. Steel lined; Bl 15; Pr MCI; St Kynoch № 20 Gastight; Wd white card/nil; Or British.

ROBERT LISLE: Queen's Hall Buildings, Derby, Derbyshire.
Business: Gunmaker.
Cartridges: Lisle's Field Cartridge; Tiger Brand; The Victa.
Example: Tiger Brand. Ga 12; Tc brown/black; De Trade Mark with a tiger; Bl 16; Pr MCI; St KN; Or British.

H. C. LITTLE & SON: Yeovil, Somerset.
Business: Gunmakers.
Cartridges: The Blackmoor Vale; The Sparkford Vale.
Example: The Sparkford Vale. Ga 12; Tc dark green/black or dark blue; De An upright cock pheasant walking in grass, L; Kn Loaded by H. C. Little & Son. Phone, Yeovil 419; Bl 16; Pr MCI; St ICI; Wd white card/nil; Or British.

CHAS F. LIVERSIDGE: Gainsborough, Lincs.
Business: Gunmaker.
Cartridges: Ejector; Special Smokeless Cartridge.
Example: Special Smokeless Cartridge. Ga 20; Tc red/black; Bl 7; Pr MCI; St EL; Or British.

LLOYD & SONS: Lewes & Horsham, Sussex.
Business: Gunmakers.
Cartridges: Champion; Champion Ejector; Improved Imps; Special Imperial Champion; Standard;
Example: Special Imperial Champion. Ga 12; Tc pine green/black; Bl 16; Pr LCI; St Lloyd Lewes 12 12 Case made in Germany; Wd pastel green/black. Lloyds x Loading x, (Inner) Smokeless 6; Or German.

C. H. LOCK: 111 Long St, Atherstone, Warwickshire.
Cartridges: Lock's Special.
Example: Lock's Special. Ga 12; Tc pale greyish green/black; De A small cock pheasant standing in scrub with tail outstretched, L; Kn Loaded specially for C. H. Lock. Entirely British made; Bl 8; Pr MCI; St ICI; Wd white card/red. Smokeless. 5; Or British.

LONDON SPORTING PARK: London.
Cartridges: Ejector.
Example: Ejector (*right*). Ga 12; One-piece full-length brass case with an inner light brown paper tube; Bl 56; Pr MCI; St Eley № 12 Ejector; Wd light jade green/black. London Sporting Park. Shot size 8 in a purple ink; Or British.

C.T. LOOK: Ely, Cambridgeshire.
Business: Ironmonger.
Cartridges: Information given was that they once sold their own brand cartridges.

LOVERIDGE & CO: 172 King's St, Reading, Berks.
Business: Ironmongers.
Cartridges: The Royal County.
Example: The Royal County. Ga 12; Tc maroon/silver; Kn Walsrode Powder. The word Cartridge is partly obscured by the high brass base; Bl 23; Pr MCI; St KB; Or British.

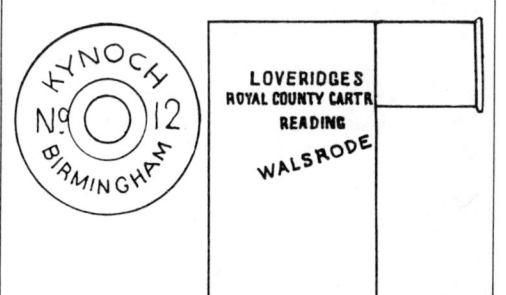

Royal County Cartridge (*below left*). Case colour: Purple with silver print. The example was taken from a fired case. Note how the wording is cut short due to the high brass head. This means that the tube was printed first.

S. LUCKES: Bridge St & Castle Green, Taunton, Somerset.
Business: Gun and ammunition merchant.
Cartridges: Taunton Demon.
Example: Taunton Demon. Ga 12; Tc eau-de-nil/blue; Kn S. Luckes. Gun and ammunition merchant. Smokeless; Bl 8; Pr MCI; St EL; Or British.

Note: This firm also had branches at Langport, Washford, Wiveliscombe and St James Foundry.

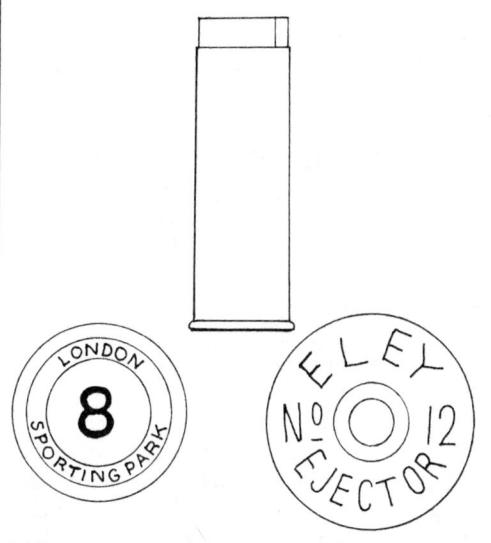

MAC – M, Mc, Mac are all treated here as Mac. The next letter in the name determines the entry position.

WILLIAM McCALL & CO: Dumfries, Dumfriesshire.
Business: Gunmakers.
Cartridges: The Border Cartridge; Eley Bros E.B.L. with own name on; All British Popular Cartridge.

McCALL & SONS: Dumfries, Dumfriesshire.
Business: Gunmakers.
Cartridges: Tally Ho.
Example: Tally Ho. Ga 12; Tc greyish green/black; De McCall & Sons written in longhand diagonally down the tube as in a signature; Kn Phone Dumfries 56; Bl 8; Pr MCI; St ICI; Wd white card/nil; Or British.
   Note: This example was taken from a scare loading that did not contain any shot.

McCOLL & FRASER: Dunfermline, Fife.
Business: Gunmakers.
Cartridges:
Example: Unnamed. Ga 12; Tc pale greyish green/black; Kn Telephone No 653; Bl 8; Pr MCI; St ICI; Or British.

McCRIRICK & SONS: 38 John Finnie St, Kilmarnock, Ayrshire.
Business: Gunmakers.
Cartridges:
Example: Unnamed. Ga 12; Tc orange/black; Kn Specially loaded by McCririck & Sons. Telephone 577; Bl 8; Pr MCI; St ICI; Wd red/black. McCririck · Kilmarnock · 5; Or British.

DUNCAN McDOUGALL: Oban, Argyll.
Cartridges: The Lorne.
Example: The Lorne. Ga 12; Tc ruby red/black; De An outlined crest of a Viking ship; Bl 8; Pr MCI; St ICI; Wd yellow/black. Smokeless · Diamond · 6; Or British.

CHARLES MACGREGOR: Kirkwall, Orkney, Orkney Islands.
Cartridges: Kynoch C. B. Cartridge Case.
Example: Kynoch C. B. Cartridge Case. Ga 12; Tc red/black; De Kynoch's (on its own) lion's head Trade Mark; Kn Kynoch C. B. Cartridge Case. Nitro Powders. Charles Macgregor. Made in Great Britain; Bl 8; Pr MCI; St KB; Wd orange buff/black. *E.C.* Powder 5; Or British.

W. McILWRAITH & CO: Elgin, Moray.
Cartridges:
Example: Unnamed. Ga 12; Tc brown quality (Witton brand)/nil; Bl 8; Pr SCI; St W. McIlwraith & Co. No 12 Elgin; Or British.
   Note: This example was taken from an unused capped case.

MACINTOSH & SONS: Cambridge, Cambs.
Cartridges: Special Smokeless Cartridge.
Example: Special Smokeless Cartridge. Ga 12; Tc primrose yellow/black; Bl 8; Pr -CI; St EL; Wd light brown/black. Eley · Loading · (Inner) Schultze 5; Or British.

ALEX MACKAY & SON: Tarbert, Argyll.
Cartridges: The Argyll.
Example: The Argyll. Ga 12; Tc orange/black; De A grouse in flight; Bl 8; Pr MCI; St ICI; Wd Not recorded; Or British.

MACKENZIE & DUNCAN: Brechin, Angus.
Cartridges: The Dunmax.
Example: The Dunmax. Ga 12; Tc orange/black; Kn Loaded by Mackenzie & Duncan; Bl 8; Pr MCI; St MGB; Wd yellow/black. Special · Smokeless · 4; Or British.

MACNAUGHTON & SONS: Edinburgh, Midlothian & Perth, Perthshire.
Business: Gunmakers.
Cartridges:
Example: Unnamed. Ga 28; Tc purple/white; Bl Not recorded; St Macnaughton. Edinburgh. Eley.; Or British.
   Though they may have had named brands, all that I have seen have been unnamed.

JOHN MACPHERSON: Later JOHN MACPHERSON & SONS: Inverness, Inverness-shire.
Business: Gunmakers.
Cartridges: The Bargate; The Barrage; The Clack; The Killer; The Royal.
Example: Unnamed. Ga 16; Tc brown tan/blue black; De A standing grouse, L; Kn Established 1887. Telegrams, Angler Inverness; Bl 11; Pr MCI; St Macpherson No 16 Inverness. Kynoch; Or British.
   Note: The example opposite was taken from an unused capped case.

SPECIALLY LOADED BY
J. MACPHERSON
GUNMAKER
INVERNESS
Established 1887
TELEGRAMS: "ANGLER. INVERNESS"

MACPHERSON Nº 16 KYNOCH INVERNESS

Case colour: Brownish orange with black print.

ROBERT MACPHERSON: Kingussie, Inverness-shire.
Cartridges: The Badenoch.

CHAS H. MALEHAM: Later MALEHAM & CO: West Bar, Sheffield & 20 Regent St, London.
Business: Gunmakers.
Cartridges: The Clay Bird; The Double Wing; The Regent; The Steeltown; The Wing.
Example: The Wing. Ga 12; Tc orange brown/black; De The Registered Trade Mark. This was a pair of wings similar to that worn by an R.A.F. pilot but with a tight shot cluster in the centre. There were seven shots in all with one forming the centre of the cluster; Kn Specially loaded by Chas H. Maleham; Bl 8; Pr MCI; St Maleham · Sheffield & London · Eley Nº 12; Wd pale pea green/black. Maleham · Sheffield · 6; Or British.

Arthur Turner in 1920 took this business on in his own name. This makes all Maleham cartridges pre-1920.

P. D. MALLOCH: 24 Scott St, Perth, Perthshire.
Cartridges: The Matchless; The Red Grouse; The Standard; The Triumph.
Example: The Matchless. Ga 12; Tc orange/black; Bl 8; Pr MCI; St ICI; Or British.

F. MANBY & BRO: Skipton, Yorks.
Cartridges: Manby's Special.
Example: Manby's Special. Ga 12; Tc orange/black; Kn Specially loaded for F. Manby & Bro; Bl 8; Pr MCI; St ICI; Wd yellow/black. Smokeless · Diamond · 5; Or British.

MANTON & CO: London & Calcutta.
Business: Gun and rifle makers.
Cartridges: Contractile; Standard Smokeless.
Example: Standard Smokeless. Ga 12; Tc buttercup yellow/black; Bl 12; Pr -CI; St Manton Nº 12 Calcutta; Loaded with a slug; Or Not known.

ALEX MARTIN incorporated with **ALEX HENRY**.
ALEX MARTIN: Glasgow, Aberdeen & Stirling.
ALEX HENRY: 12 Andrew St, Edinburgh, Midlothian.
Business: Gunmakers.
Cartridges: The AGE; The Calendonia; The Club; High Velocity; The Scotia; The Stirling; The Thistle; The Thistle High Velocity; The Velm.
Example one: The AGE. Ga 12; Tc orange/black; De The AGE Registered Trade Mark (A for Aberdeen, G for Glasgow and E for Edinburgh); Bl 8; Pr MCI; St ICI; Wd orange/black. Martin · Glasgow & Aberdeen · E6C; Or British.

Note: This brand was also loaded in Ga 20 and Ga 16.
Example two: Martin's Scotia. Ga 12; Tc gastight brick red/black; Kn 20 Exchange Square, Glasgow. 22 Frederick Street, Edinburgh. 25 Bridge Street, Aberdeen. 2 Friars Street, Stirling; Steel liner; Bl 16; Pr MCI; St ICI; Wd brown card/black (Cluster type) 7; Or British.
Example three: The Velm. Ga 12; Tc grey/black; Kn Resilient, Accelerated. Loaded by Alex Martin Glasgow, Aberdeen and Stirling. Incorporated with Alex Henry & Co. Edinburgh. Established over 130 years; Bl 8; Pr MCI; St ICI; Wd white card and pen inked with the word Ball. Square turnover; Or British.

J. F. MASON: Eynsham Hall, Oxon.
Business: Private estate.
Cartridges: Ejector.
Example: Ejector. Ga 12; Two-piece brass case with an inner paper tube; Pr MCI; St J. F. Mason. Eynsham Hall. № 12 Eley; Or British.
Note: The details of this example are taken from woodland remains.

J. MATHER & CO: Newark & Southwell, Notts.
Business: Ironmongers.
Cartridges: The Britannia Cartridge.
Example: The Britannia Cartridge. Ga 12; Tc vermilion/black; De The Britannia crest; Kn Mather's name also on the paper tube; Bl 8; St Jas. R. Watson & Co. London; Or Not known.

MATTERSON, HUXLEY & WATSON: Bishop St, Coventry, Warwickshire (W. Midlands).
Cartridges: The only information I have is that a cartridge has been seen with their name on it.

JAMES MATTHEWS: 42 Ballymoney St, Ballymena, Antrim, Northern Ireland.
Business: Dealer in hardware and guns.
Cartridges: Hawk; Kingfisher; Swift; Wizard. This firm was founded in 1906. Information that I was once given is Hawk a brown foreign case. Kingfisher a blue case. Swift a yellow case with a large cap. Wizard a green case. Cartridges seen had the St SS.

MAWER & SAUNDERS: The Square, Market Harborough, Leics.
Business: Ironmongers.
Cartridges: I was told that they once had their name on cartridges and might even have loaded some.

N. MELLARD: Denbigh, Denbighs (Clwyd).
Business: Ironmonger
Cartridges:
Example: Unnamed. Ga 12; Tc blue quality/white; De A standing cock pheasant; Bl 10; Pr -CI; St KB; Or British.

W. METCALF: Richmond & Shute Rd, Catterick Camp, Yorks.
Cartridges: The Special.
Example: Unnamed. Ga 12; Tc dark blue/not recorded; De A running hare; Bl 8; Pr MCI; St ICI; Or British.

G. M. MICHIE & CO: Stirling, Stirlingshire.
Cartridges: Michie's Unequalled.
Example: Michie's Unequalled. Ga 20; Tc dark green/black; Bl 10; Pr SCI; St ELG; Or British:
Note: This example is taken from an unused capped case.

MIDLAND GUN CO: Demon Gun Works, Bath St, Birmingham 4 (W. Midlands).
Business: Gun and cartridge makers.
Cartridges: Having had access to an old catalogue, I am able to give some details with their brand names.
The Best Of All (Loaded in Ga 12. 65. 70 and 75 mm; Ga. 16, 65 and 70 mm; Ga 20. 65 mm). Demon (Ga 12. 65 mm. Named after their gun works). The Double Demon (Ga 12. 70 mm. Listed as a pigeon cartridge it was steel lined with Bl 25). The Edward (Ga 12. 65 mm. Tc grey/dark blue or black. Smokeless Diamond Powder. Marketed in 1937). Ejector (Ga's 12 and 16. 65 and 70 mm. Full-length brass cases). The Imp (Tc red/black). The Jubilee (Ga, 12. 65 mm. Tc pale green/dark blue Bl 8. Marketed in 1937). The Keeper. Also The Keeper H. V. (Ga 12. 65, 70 and 75 mm; Ga 16. 65 and 70 mm; Ga 20. 65 mm). Perfection Smokeless (Ga's 12, 16 and 20. 65, 70 and 75 mm. Gastight steel lined with Bl 25). Perfect Smokeless; The Rabbit Special Smokeless; the Record; Record Brand; Sudden Death: (Ga's 12, 16 and 20. 65, 70 and 75 mm. Gastight steel lined with Bl 16).
Example: The Rabbit Special Smokeless. Ga, 12; Tc yellow/black; De A bowled-over rabbit being peppered with shot; Kn Hand loaded by the Midland Gun Co. Birmingham Eng; Bl 8; Pr MCI; St SS; Or Not known. This firm was founded in 1888 at 77 Bath Street and it moved to Vesey Street in 1902. They also sold pinfire cartridges and a very early catalogue had 15-bore cartridges added in with an ink pen. The Midland Gun Company also loaded ·410 cartridges in short and long. They were also known to have loaded cartridges for other firms by special order. Their telegraphic address was 'Rifles Birmingham' and their telephone number was Central 1254. During December 1956 they were absorbed into Parker Hale.

MILBURN & SON: 5–7 High Cross St, Brampton, Cumberland (Cumbria).
Business: Gunmakers
Cartridges: The Don; The Milburn; M.S.B. (Milburn & Son, Brampton); The Noxall; The Rex.
Example: The Noxall. Ga 12; Tc eau-de-nil/dark blue; Kn Loaded by Milburn & Son, Brampton; Bl 8; Pr MCI; St ICI; Wd white card/red with the print forming a background for white lettering. Milburn & Son · Brampton · 8; Or British.

R. MILLETT: Ilminster, Somerset.
Cartridges: Their name has been seen printed in black on a white card top wad.

MINTO: Wigton, Cumberland (Cumbria).
Cartridges: Though this name has been given to me, I have no cartridge information.

W. H. MONK: Also as HENRY MONK: Chester, Cheshire.
Business: Gunmakers.
Cartridges: The Imperial; Pegamoid; The Popular; The Royal.
Example one: Unnamed. Ga 20: Tc red/black; De Trade Mark (Rear view of a rabbit in front of a shock of corn with WHM monogram); Kn Loaded and guaranteed by W. H. Monk, Chester; Bl 10; Pr MCI; St ELG; Or British.
Example two: Unnamed. Ga 12; Tc olive green/black; De Trade Mark (as on Example one); Kn, Henry Monk, Gunmaker, Chester. Telephone No. 988; Bl 16; Pr MCI; St ICI; Wd white card/red. Smokeless 5; Or British.
  Monks loaded many similar cartridges without brand names in different-coloured cases in Ga's 12, 16 and 20.

W. F. MOODY: 13 Church St, Romsey, Hants.
Business: Gunmaker and cutler.
Cartridges: Moody's Special Waterproof Non Corrosive Cartridge.
Example: The above. Ga 12; Tc maroon/silver; De A walking hen pheasant, L; Kn, British hand loaded; Bl 8; Pr MCI; St SS; Wd cream/black. Special Smokeless 4; Or Foreign.
  His name has also been seen on the top wad that was loaded into a white-tubed pinfire with the St ICI.

T. H. MOOR: South Molton, Devon & Exford, Somerset.
Business: Ironmonger.
Cartridges: The Molton; Special Rabbit.

MOORE & GREY:London
Business: Gunmakers.
Cartridges:
Example: Unnamed pinfire. Ga 12; Tc green/nil; Bl 6; St (Reversed) Moore & Grey. London. 12; Wd plain; Or Not known.

MORREYS: Holmes Chapel, Cheshire.
Cartridges: Mooreys Special.
Example: Mooreys Special. Ga 12; Tc red/black; Kn Specially loaded for Mooreys. Phone 25; Bl 8; Pr LCI; St SS; Wd pink card/nil; Or Not known.

P. MORRIS & SON: Hereford, Herefordshire.
Business: Gunsmiths and ironmongers.
Cartridges: The Hereford; The Imperial.
Example: The Imperial. Ga 12; Tc orange/black; Kn Smokeless Cartridge; Bl 8; Pr MCI; St KN; Wd Not recorded; Or British.

MORROW & CO: Halifax & Harrogate, Yorks.
Cartridges: Challenge Cartridge.

MORTIMER: Barnstaple, Devon.
Cartridges: The Club.

MORTIMER & SON: Edinburgh, Midlothian.
Business: Gunmakers.
Cartridges:
Example: Unnamed. Ga 12; Tc middle blue/dark blue; Kn Established 1720. Telephone 26761; Bl 8; Pr MCI; St ICI; Wd deep blue/black. Mortimer & Son · Edinburgh · 6; Or British.
  Mortimer & Son incorporated the business of Joseph Harkom & Son.

MULLER & CO: Horseshoe Yard, Mount St, London. Also at Winchmore Hill, Middx.
Business: Cartridge agents.
Cartridges: Clermonite; Mullerite; Negro.
  This firm set up business in Mount Street in 1901 to market Mullerite-brand cartridges in England. They moved to Winchmore Hill in 1903 only to close down in 1905. See Martin Pulverman & Co. The company then became the agents for the Mullerite cartridges.

THE MULLERITE CARTRIDGE WORKS: 59 Bath St, and St Mary's Row, Birmingham.
Business: Cartridge makers.
Cartridges: The Ace; The Ace Long Range; The British Champion; The Champion Smokeless; The Champion Smokeless Heavy Load; Fourtenner Long; Fourtenner Short; General Service; Green Seal; Grey Seal; Heyman Smokeless; Red Seal; Silver Ray; Smokeless; Special Clayking; Yellow Seal.
Example one: Silver Ray. Also Grey Seal. Ga 12; Tc black/silver; De Martin Pulverman MPL monogram; Kn British loaded; Bl 26; Pr LBN; St Special 12 12 Gastight; Or Not known. Note: This cartridge named Silver Ray also carried the name Grey Seal. Some of their cartridges went by the Seal name only.
Example two: Yellow Seal. Ga 16; Tc red/black; De Martin Pulverman monogram; Kn British loaded. Smokeless; Bl 9; Pr LCI; St Smokeless 16 16 Gastight; Wd Dark red/black. Special · Smokeless · 5; Or Not known.
Example three: Yellow Seal. Ga 12; Tc yellow/black; De Martin Pulverman monogram. The name Mullerite in long hand; Kn Smokeless. Anti-corrosive foreign cap and case. Loaded in Great Britain; Bl 8; Pr LCI; St SS; Wd yellow/red. Mullerite · Clermonite · 4; Or Foreign.

The Mullerite Cartridge Works were established in Birmingham in 1922. These works were known to have loaded cartridges for other firms to order, often using yellow-coloured cased Mullerite cartridges with extra tube printing added. The Heyman Smokeless cartridge name was seen on an old Mullerite cartridge box label wrap and the cartridge may have had a peacock with its tail fanned out on the tube as this was portrayed on that label wrap.

T. W. MURRAY & CO: Cork, Republic of Ireland.
Cartridges: The Speedwell Smokeless; The Wildfowler.
Example: The Wildfowler. Ga 12; Tc orange/black; De A duck being peppered with shot within a circle; Kn Loaded with Smokeless Diamond Powder; Bl 8; Pr MCI; St ICI; Wd yellow/black. Smokeless · Diamond · IX; Or British.

NATIONAL ARMS & AMMUNITION CO: (N. A. & A. CO.): Perry Bar, Birmingham.
Business: Cartridge manufacturers, ammunition and gunmakers.
Cartridges: Ejector; Express.
Example: Ejector. Ga, 12. One-piece thin brass case with thin rim, having no separate base; Bl 65; Pr MCN; St N. A. & A Cº Nº 12 Patent; Or, British. Note: The example is taken from an unused capped case.

This firm was in business from 1872 until 1882. Other addresses were 77 Bath Street, Birmingham, and Montgomery Street, Sparkbrook. Their thin brass cases have also been seen in Ga's 14 and 16.

T. NAUGHTON & SONS: Galway, Republic of Ireland.
Business: Sports dealers.
Cartridges: The Blazer; Connaught.
Example: Connaught. Ga 12; Tc dark green/black; Kn Special loading. Telephone 63. A product of Irish Metal Industries L^{td}. Manufactured under licence; Bl 16; Pr MCI; St EK; Wd red card/black. (A circle only with the shot size) 6; Or Irish.

C. NAYLOR: Sheffield, Yorks.
Cartridges: Naylor's Cannot be Beaten.
Example: Naylor's Cannot be Beaten. Ga 12; Tc orange/black; De Cock pheasant standing in grass with tail horizontal, L; Kn Only writing is the name of the cartridge; Bl 10; Pr MCI; St C. Naylor Nº 12 Sheffield; Wd orange/black. C. Naylor · Sheffield · 6; Or Not known.

Naylor's Cannot Be Beaten (*opposite below*). Case colour: Orange with black print. Wad colour as illustrated: Orange with black print.

J. V. NEEDHAM: 20A Temple St. Also Damascus Works, Loveday St, Birmingham.
Cartridges: Uneedem.
Example: Uneedem. Ga 12; Tc mauve/black; De Rabbit bounding in the run with all legs outstretched, R; Kn Double printed ring below the turnover. Smokeless; Bl 8; Pr LCI; St J. V. Needham *12*; Or Not known. Note: The example is taken from an unused capped case.

FRANCIS NELSON: Sligo, Republic of Ireland.
Cartridges: The Reliable.
Example: The Reliable. Ga 12; Tc brown/black; De A low-flying bird; Kn Loaded with Nobel's Empire Powder; Bl 8; Pr Not recorded; St NG; Or Scottish.

A. NESTOR: Limerick, Republic of Ireland.
Cartridges: Ejector.
Example: Ejector. Ga 12; Two-piece brass Kynoch Grouse Ejector with a brown inner paper tube; Bl 10 + 60; Pr MCI; St (Outer) Kynoch № 12 Patent. (Inner) № 2090 Grouse Ejector; Wd white card/red. A. Nestor · Limerick · 2; Or British.

THE NEW EXPLOSIVES CO: Stowmarket. Also at 62 London Wall, E. C.

A monogram of The New Explosives Co.

Business: Powder and ammunition manufacturers.
Cartridges: Fourten; The Green Rival; The Neco; The N.E. Powder; The Premier; The Primrose Powder; The Red Rival; Sixteen Bore; Twenty Bore.
Example: The Neco. Ga 12; Tc greyish green/black; De The New Explosives Co's large round monogram crest; Kn Smokeless cartridge; Bl 8; Pr MCI; St EL; Wd yellow/black (Shot size only) 3; Or British.

Cartridges made by The New Explosives Company date from 1907 until loading ceased in 1920.

Case colour: Greyish green with black print. Wad colour: Caprice yellow with black print.

GEORGE NEWHAM: Also **NEWHAM & CO**:
29 Commercial Rd, Lamport, Portsmouth, Hants.
Business: Gunmakers.
Cartridges: The Champion; Ejector; The Keepers Cartridge; Pegamoid; The Special Game; Kynoch's Deep Shell Cartridge Case.
Example: The Special Game. Ga 12; Tc cream/black; De A small cock pheasant walking through scrub with tail horizontal, L; Kn Specially loaded by Newham Company; Bl 8; Pr MCI; St ICI; Or British.

NEWLAND & STIDOLPH: Stratford-upon-Avon, Warwickshire.
Business: Ironmongers.
Cartridges:
Example: Unnamed Ga. 12; Tc brown quality/black; De A crouching quail Fl; Bl 8; Pr MCI; St Newland & Stidolph. №12 Eley; Wd white card/black (Cluster type) Six; Or British.

NEW NORMAL AMMUNITION CO: Hendon, London, N.W.4.
Business: Cartridge loading and merchants.
Cartridges: The details given below were taken from an old advertisement.
Gastight (Ga's 12, 16 and 20); The Hendon (Ga's 12 and 16); The Nimrod (Loaded with Walstrode Power in Ga's 12, 16, 20 and 28); The Normalis (Ga ·410 in 50 and 65 mm case length); Special Twenty.
Example: The Hendon. Ga 16; Tc orange tan/black; Kn Smokeless cartridge. Bl 7; Pr MCI; St (Outer) Smokeless 16 16 Gastight. (Inner) Foreign made case; Wd white card/red. Special · Smokeless· 5; Or Foreign. Note: The Special Twenty was either an early or later cartridge as it was not mentioned on the advertisement referred to above.

T. NEWTON: 48 King St West, Manchester, Lancs (Gt. Manchester).
Business: Gunmaker.
Cartridges: The Lightning; Pegamoid Brand Paper; The Smokeless Cartridge; Newton's G.P.
Example: Unnamed. Ga 12; Tc brown/black; Kn Tel. Blackfriars 5674. Water resisting; Bl 16; Pr MCI; St ICI. Wd white card/dark blue. T. Newton · Manchester · 5½; Or British.

J. H. NICHOLAS: Thirsk, Yorks.
Business: Ironmonger.
Cartridges: The Express.
Example: The Express. Ga 12; Tc black/silver; Bl 10; Pr LCI; St SS. No other details recorded.

J. O. & R. W. NICOLL: Aberfeldy, Perthshire.
Cartridges:
Example: Unnamed or name not known. Ga 12; Tc blue/black; Kn Hand loaded. Shooter's best hand made water tights made to measure; Bl 8; Pr MCI; St EN; Or British.

NIGHTINGALE & SON: Salisbury, Wilts.
Cartridges: The Avon; The Moonraker.
Example: The Moonraker. Ga 12; Tc orange/black; De A standing partridge, L; Kn Smokeless. British hand loaded; Bl 8; Pr MCI; St SS; Wd orange/black. W. Nightingale · Salisbury 6; Or Foreign.

The Moonraker. Case colour: Deep orange with black print. Wad colour as illustrated: Dark orange with black print.

THE NITROKOL POWDER CO: London.
Cartridges: The Redskin; The Rover.
Example: The Redskin. Ga 12; Tc flame red/black; Kn Best British loading. Smokeless Cartridge. Made France (These two words in small print running around the tube so that when loaded they would be rolled in with the turnover); Bl 7; Pr MCI (Copper insert cup); St Nitrokol № 12 London; Or French. Note: The Rover and the Redskin are identical except for the brand name. Only ready capped unused cases have so far been seen.

NIXON & NAUGHTON: Newark, Notts.
Business: Gunmakers.
Cartridges: It is believed that this firm had its name on cartridges.

NOBEL EXPLOSIVES CO: Glasgow, Scotland. Also London, England.
Business: Gunpowder and ammunition manufacturers.
Cartridges: Ajax; Clyde; Corio (See I.C.I.A.N.Z. in An Australian Collection towards the rear of this book); Ejector; Empire; Eureka; Gas-tight; Kardax; Kingsway; National; Nile; Nitro; Noneka; Orion; Parvo; Pegamoid; Regent; Special Primrose; Sporting Ballistite; Target; Valeka.
Example One: National. Ga 12; Tc pale green/black; De Two Union Jack flags with the word Nobel's centralized within; Bl 8; Pr MCI; St NG; Or Scottish.
Example Two: Noneka. Ga 20; Tc purple/silver white; Bl 16; Pr MCI; St NG. Wd mauve/black. Nobel's · Loading · 6; Or Scottish.

This famous firm was first founded during 1871 as the British Dynamite Co. They later became famous for their Sporting Ballistite gunpowder marketed from 1889 onwards. In 1907 they acquired the business of Frederick Joyce & Co, the plant and patents being used for their own supply of cartridge cases. By examining both Joyce and Nobel Ejector brass cartridge cases the similarity can be clearly seen. It is believed that prior to this takeover, some cases were imported from America. Many of this firm's brand cartridges took their names from the Nobel powder that was loaded into them. Cartridges were also loaded to order for other firms. Mr Eric Wastie, the well-known cartridge collector in the Oxford area, did once find two interesting old cartridge heads. These had the headstamping Nobel London as opposed to the Nobel Glasgow that is so often seen. But where, and for how long in London, they might have been made would be interesting to know.

NOBEL INDUSTRIES LIMITED: Witton, Birmingham & London.
Business: Ammunition manufacturers.
Cartridges: I am not certain about all the brand names used, but I am listing those that I think most likely. Following these I have given three examples each with a different headstamping. Acme; Bonax; Clyde; Deep Shell; Ejector; Eley ⅝" Brass; Fourlong; Fourten; Gastight; Gastight Quality Case; Grand Prix; Grand Prix Quality Case; Kardax; Majestic; Nitrone; Parvo; Pegamoid; Primax; Smokeless Cartridge; Twenty Gauge; Two-Inch; Velocity; Westminster; Winchester Cannon; Yeoman; Zenith.
Example One: Gastight. Ga 32; Tc brown/black; De The EBL shield Trade Mark; Steel lined; Bl 12; Pr MCI; St. Eley N.I. London. 32; Or British.
Example Two: Majestic. Ga 12; Tc orange/black; De Lion's head Trade Mark; Bl 8; Pr MCI; St EN; Or British.
Example Three: Primax. Ga 12; Tc dark green/black; De Lion's head in circles Trade Mark; Steel lined; Bl 16; Pr MCI; St KN; Or British.

This firm was formed towards the end of 1920. It was in the November of 1918 that the large mergers brought about the birth of Explosives Trades Limited, only to be renamed Nobel Industries two years later. It was with this name that the firm manufactured and marketed their cartridges for the next five to six years, then to change its name yet again to I.C.I. Metal Division (Eley–Kynoch) Ltd, when Imperial Chemical Industries took over the controlling interests. It was at this time (circa 1926) that the headstamps Eley Nobel and Kynoch Nobel gave way to the long-run Eley-Kynoch I.C.I. Apart from manufacturing their own brands, the firm continued in its normal practice of loading a large percentage of its cartridges for the many smaller firms in the British Isles and also for the export market.

Although many small firms have been absorbed into this large organization, four firms of major importance can be singled out. These were Eley Brothers, Kynoch & Co, Nobel Explosives Co and the Schultze Gunpowder Co. It was because of this that many of their brand names were retained on the cartridges, although the largest portion of them were those from Eley's. Cartridges were also manufactured under licence in Ireland and Australia. Down under many of the same brand names that were marketed at home were also made and these mingled with some of the Australian brand names. (See An Australian Collection.) Joyce was also remembered in the sale of percussion caps.

At the time of the mergers many of the cartridges that had already been made would have been sold off by the new firm. I have seen both Eley and Kynoch Gastights with the N.I. headstamps. While talking of headstamps, early stampings had the letters N.I. on them following the practice of E.T.L. (See Explosives Trades Limited in this list.) The later stamps had the names Eley Nobel and Kynoch Nobel, so honouring the names of Eley, Kynoch and Nobel respectively. Most shops in the British Isles selling their own brands of cartridges produced by the Witton factory, also sold the brands made by the parent firm.

NORMAL IMPROVED AMMUNITION CO: Hendon, London. N.W.4.
Business: Manufacturers.
Cartridges: ·410 Bore Cartridge; Hendon; Keepers Normal; Light Blue; Normal; Normal Midget; Pegamoid; Pigeon Cases; Super Nimrod.
Example: Light Blue. Ga 12; Tc light blue/black; St Normal № 12 London. No other details recorded.

This firm seems to be that of the Normal Powder Co in an interim period prior to becoming the New Normal Ammunition Company. I have listed above brand names to the best of my knowledge.

NORMAL POWDER CO: Hendon, London. N.W.4.
Business: Gunpowder manufacturers and cartridge loaders.
Cartridges: Ejector; Nimrod; Wasters; Waterproof.
Example: Wasters. Ga 12; Tc waxed signal red/black; Kn The word Wasters with print running towards the closure is the only print on the tube; Bl 11; Two or three rings run around the side of the brass; Pr 5 mm diam cap fitted into a 12 mm insert cup; St's Normal № 12 Waterproof, Normal № 12 W.R.A. Co., Normal № 12 Nimrod; Wd white card/red. Nimrod. Normal Powder 8; Or U.S.A.

Note: There is a possibility that the name Wasters may have been that of another firm for whom the Normal Powder Co had loaded.

I am still very much in the dark on the history of this firm. They obviously became the New Normal Ammunition Co. The questions that need answering are when, why and what was their address during their time in business? Also their period of activity? Many of the cartridges that have turned up have had no brand names and often their tubes refer to No blow-backs and No gun headaches. As you will see by the example, I have described a cartridge for which I have seen three different headstampings. The head with W.R.A. Co. on it means that it was imported from the U.S.A. These letters stand for the Winchester Repeating Arms Company of New Haven, Connecticut. It is not known when this firm produced their first shotgun cartridges or shells, but they were known to have been waterproofing them by 1884. The letters W.R.A. Co. were used on a lot of their early stampings and certainly prior to the word Winchester. Frank H. Steward illustrates in his book *Shotgun Shells* (an American publication) a similar type of head, the Winchester № 12 Normal. This makes me wonder if the Normal Powder Company started out as an offspring from the W.R.A. Co., later to become Winchester and Winchester Western.

NORMAN & SONS: Woodbridge & Framlingham, Suffolk.
Business: Gunmakers.
Cartridges: The Gastight; The Service; The Special; The Standard.
Example: The Standard. Ga 12; Tc middle blue/black; Kn Loaded by Norman & Sons. Established 1870. Telephones Woodbridge 144 and Framlingham 90; Steel liner; Bl 16; Pr MCI; St ICI; Or British.

J. H. B. NORTH & SONS: Stamford, Lincs & Peterborough, Northants (Cambs).
Cartridges: North's Universal.

ODELL BROTHERS: 13 High St, Newport Pagnell, Bucks (Milton Keynes).
Business: Ironmongers and merchants.
Cartridges: This firm was established over 200 years ago and it has since changed hands. I was told by the Odell Brothers that many years ago they did load and sell their own cartridges. Although Odell's let me have some interesting old cartridges from other ironmonger firms they never managed to find one of their own. Perhaps they even loaded for other ironmongers.

OLBYS: Ashford, Canterbury, Folkestone, Margate & Ramsgate, Kent.
Cartridges: Olbys Cantium.
Example: Olbys Cantium. Gas 12; Tc signal red/black; De A riding crop; Kn Special smokeless; Bl 7; Pr LCI; St SS; Wd red/black. Special · Smokeless · 5; Or Not known. Note: This cartridge has the look of a Midland Gun Company loading.

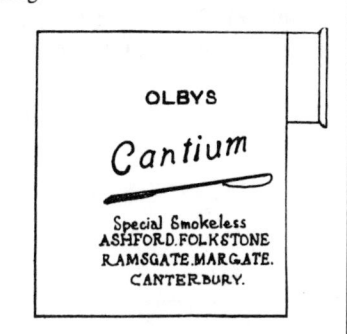

Cantium. Case colour: Signal red with black print. Wad colour as illustrated: Signal red with black print.

HERBERT O'LEE: Bishop's Stortford, Herts.
Cartridges: The Sharpshooter.
Example: The Sharpshooter Ga 12; Tc salmon pink/black; Kn Smokeless cartridge; Bl 8; Pr Not recorded; St MGB; Or British.

OLIVER & CO: Hull, Yorks (Humberside)
Business: Gunmakers
Cartridges: The Estate.
Example: The Estate. Ga 12; Tc green/black; Kn Smokeless; Bl 8; St № 12; Or Not known. No other details recorded.

J. P. OSBORN: The Golden Padlock, Daventry, Northants.
Business: Ammunition and cartridge sales.
Cartridges: The Danatre.
Example: The Danatre. Ga 12; Tc ruby red/black; Kn Ammunition and cartridge depot; Bl 8; Pr LCI; St SS. Wd red/black. Special · Smokeless · 6; Or Foreign.

WALTER OTTON: Exeter, Devon.
Business: Ironmonger.
Cartridges: The Express; The Long Tom.
Example: The Long Tom. Ga 12; Tc purple/yellow Bl 8; Pr MCI; St KB; Wd white card/black. (Cluster type) 5; Or British.

T. PAGE-WOOD: Bristol (Avon).
Business: Gunmakers and cartridge loaders.
Cartridges: Anti-Recoil Cartridge; Anti-Recoil Economic Cartridge; The Bristol; The Climax Cartridge; The Double Crimp; The Imperial Crown; The Page-Wood DS; Page-Wood's Shield Cartridge; The Park Row; Special ·410; The Wildfowler.
Example One: Page-Wood's Shield Cartridges. Ga 12; Tc Pale greyish green/black; De A large black shield with the name and a picture of the cartridge on it; Kn Manufactured by T. Page-Wood Ltd; Bl 8; Pr MCI; St ICI; Their own patent crimp-roll closure; Wd red/black. Patent 6; Or British.
Example Two: The Imperial Crown. Ga 12; Tc red/black; De A crown replacing the Page-Wood shield, Bl 16; St T Page-Wood Bristol. No other details recorded.
Example Three: The Climax Cartridge. Ga 12; Tc eau-de-nil/dark blue; Kn For rough shooting; Bl 8; Pr MCI; St ICI; Normal turnover; Or British.

JAMES PAIN & SONS: Salisbury, Wilts & London.
Business: Pyrotechnics and signal cartridge manufacturers.
Cartridges: Bird Scaring Cartridge.
Example: Unnamed. (Ejector type). Ga, 4; 100 mm length brass case with a dark green inner paper tube; Bl 6 + 94; St J. P. & S. IV; Or Not known.

G. PALMER: 29 High St, Sittingbourne, Kent.
Cartridges: The Champion.
Example: The Champion. Ga 12; Tc ruby red/black; Bl 8; Pr MCI; St KB; Wd white card/black. (Cluster type) 6; Or British.

PALMER & SONS & CO: Barnet, Herts (Gt. London).
Cartridges: Rocketer.
Example: Rocketer. Ga 12; Tc flesh pink/black; Bl 10; Pr -CI; St ELG; Or British.

W. P. JONES: Also as W. PALMER JONES (GUNS) LTD: 25 Whittall St, Birmingham.
Cartridges: The Accuratus; The Priority.
Example: The Accuratus. Ga 12; Tc primrose yellow/black; Kn The tube printing runs around the cartridge and is read when the cartridge is standing on its base. Specially loaded by W. P. Jones; Bl 10; Pr SCI; St W. P. Jones. Birmm. № 12 Eley; Wd orange card/purple. W. P. Jones. B'ham 4; Or British.

W. R. PAPE: 21 Collingwood St, Newcastle-upon-Tyne & Sunderland (Tyne & Wear).
Business: Gunmaker.
Cartridges: The Beryl; The Heather; The Ranger Smokeless, The Setter.

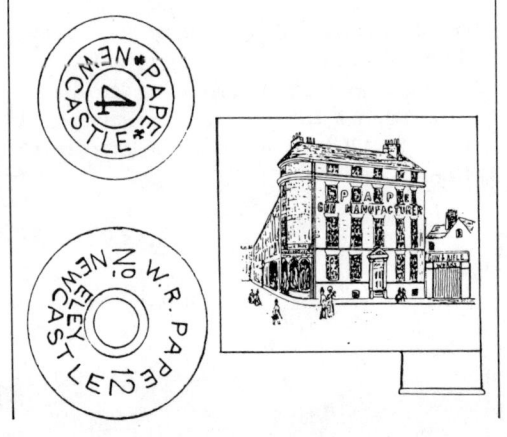

Example: Unnamed. (*below left*). Ga 12; Tc brownish orange/black; An engraved-type drawing of the street showing the gunworks and shop with guns in the window. This picture takes up the whole of the tube and is viewed while the cartridge is standing on its base; Bl 10; Pr MCI; St W. R. Pape № 12 Newcastle Eley; Wd pink card/black. Paper * Newcastle * 4; Or British.

William Rochester Pape in his time marketed very many colourful and decorative cartridges, often without any brand names. Some of these illustrated game or guns on their paper tubes. The Beryl Cartridge was named after his wife. Mr W. R. Pape died during 1923.

The Beryl. Case colour: Greyish green with black print.

Left: Case colour: Dull orange with black print. Wad colour as illustrated: Deep pink with black print.

PARAGON GUNS: 43 Ann St, Belfast, Northern Ireland.
Business: Gun specialists.
Cartridges: The Crown; The Invincible; Paragon Special; Special.
Example: "Paragon Special". Ga 12; Tc blue/black; De A standing cock pheasant; Bl 8; Pr Not recorded; St SS; Wd white card/black. Special · Smokeless · 5½; Or Not known.

TOM PARKINSON: Ulverston, Lancs (Cumbria).
Cartridges:
Unnamed. Example: Ga 12; Tc ruby red/black; Bl 8; Pr MCI; St KB; Or British.

C. PARSONS: Nuneaton, Hinkley & Coventry (Midlands).
Business: Agricultural and hardware merchants.
Cartridges: Special Loading.
Example: Special Loading. Ga 12; Tc turquoise/dark blue; De A rabbit bounding over a small foxglove plant, FL; Bl 8; Pr MCI; St, C. Parsons Nº 12 Nuneaton; Wd white card/black. (Shot size only within a circle) 6; Or Not known, believed that it may have been British.

PARSONS, SHERWIN & CO: Nuneaton, Hinkley & Coventry (Midlands).
Business: Agricultural and hardware merchants.
Cartridges: Special Loading.
Example: Special Loading. Ga 12; Tc orange/black; De A cock pheasant standing in grass with tail nearly horizontal, L; Bl 10; Pr MCI; St Parsons, Sherwin & Co Ltd. ·No12· ; Wd orange tan/black. Parsons Sherwin & Cº· (Inner) Kynoch Smokeless 6; Or British.

When their old buildings in Nuneaton were knocked down several years ago, a room at the rear that looked as though it had been locked up since the Great War revealed its secrets. Covered in thick dust was an old loading machine with hundreds of unused ready-capped cases printed for C. Parsons and also for Parsons, Sherwin & Company. I do not know when Parsons met Sherwin or when they decided to go public, but these cartridges were loaded at Nuneaton and very many years ago. Although I have never seen a more recent round with their name on, it is quite possible that cartridges may have been loaded for them at a much later time.

PATSTONE & COX: Southampton Winchester, Hants.
Business: Gunmakers.
Cartridges: The Pheasant; The Precision.
Example: The Precision. Ga 12; Tc middle green/gold; Kn Established over a century; Bl 16; Pr S-I; St SS; Or Foreign.

PATSTONE & SON: 28 High St, Southampton. Also at Winchester, Hants.
Business: Gunmakers.
Cartridges: The Precision; The Reliable; The Renown.
Example: Unnamed. Ga 16; Tc burgundy/silver; De A hen pheasant walking on firpins with tail high, L. Paris medals dated 1878 and London medals dated 1885; Kn Established over a century; Bl 7; Pr MBI; St Smokeless 16 16 Gastight; Wd white card/red. Special · Smokeless · 6; Or Foreign.

J. C. PATTERSON: Lisburn, County Down, Northern Ireland.
Cartridges: The Nailer.

JOSEPH PEACE: Darlington, Co. Durham.
Business: Gunmaker and sports outfitters.
Cartridges:
Example: Eley G.P. Case. Ga 12; Tc orange/black; De The EBL shield Trade Mark; Kn J. Peace's name was on the top wad only; Bl 8; Pr MCI; St ICI; Wd deep blue/black. J. Peace Ltd · Darlington · 6; Or British.

PECK: Ely, Cambs.
Cartridges:
Example: Unnamed or name not known. Ga 12; Tc maroon -; Kn The example in a very poor condition; Bl 9; Pr -CI; St Peck Nº 12 Ely; Or Not known.

PERRINS & SON: Worcester, Worcs.
Cartridges: Ejector.
Example: Ejector. Ga 12; An all-brass case with a double head; St, Perrins & Son, Worcester. Eley; Or British. No other details recorded.

S. PERROTT: Kingsbridge, Devon.
Business: Sadler and ironmonger.
Cartridges: Gastight Cartridge.
Example: Gastight Cartridge. Ga 14; Tc light orange paper tube with the wording 'S. Perrott. Kingsbridge.' in a purple ink. The following tube printing in black – Patent Gastight Cartridge charged with Schultze Gunpowder; Bl 10; Pr MCI; St KB; Or British.

Note: Not very many small firms had their names placed on 14-bore cartridges.

PHILLIPS & POWIS: 34 & 37 West St, Reading, Berks.
Business: Cycle agents.
Cartridges: The Pheasant.
Example: The Pheasant. Ga 12; Tc dark red/black; De A cock pheasant shown walking without any ground and a short low-positioned tail, L; Kn Smokeless. Hand loaded; Bl 8; Pr MCI; St Smokeless 12 12 Gastight; Wd Signal red/black. Hand Loaded + 5; Or Not known, believed to be foreign. Note: This cartridge looks as though it may have been loaded for them by the Midland Gun Company.

The Reading Borough Library gave a different address for this firm and referred to them as motorcar and aviation agents. They did later go into business at Woodley Aerodrome with Miles Aircraft. They then became well known for their civil range of Hawk aircraft and their military trainers, Magisters, Masters and Martinets.

The Pheasant. Case colour: Dark or monarch red with black print. Wad colour as illustrated: Signal red with black print. Believed to have been a Midland Gun Company's loading. This firm later dealt in motors and then aircraft when they joined forces with Miles on Woodley Aerodrome.

PINDERS: Market Place, Salisbury, Wilts.
Business: Ironmongers.
Cartridges: The Deadshot.
Example: The Deadshot. Ga 12; Tc green/black; Kn Loaded for them by Frank Dyke of London. No other details recorded.

C. PINDERS: Basingstoke, Hants.
Cartridges: This firm's name has been seen on Ga 12 cartridge heads of pre-First World War vintage dug up in a garden.

PLUMBERS: Great Yarmouth, Norfolk.
Business: Known to have sold guns and ammunition.
Cartridges: Norfolk High Velocity Load; The Original Norfolk.
Example: The Original Norfolk. Ga 12; Tc waxed blue/black; Bl 8; Pr L--; St Plumbers. Gt. Yarmouth. No other details recorded.

PNEUMATIC CARTRIDGE CO: 61–67 Albert St, Edinburgh, Midlothian.
Business: Cartridge loaders.
Cartridges: Pneumatic No 1; Pneumatic No 2.
Example: Pneumatic No 2. Ga 12; Tc red/black; Kn Standard-quality cartridge; Bl 8; St Pneumatic Cartridge Co. Edinburgh. Eley. No other details recorded.

PNEUMATIC CARTRIDGE CO: 96–98 Holyrood Rd, Edinburgh 8. Midlothian.
Business: Cartridge experts.
Cartridges: Ejector; Pncuma, Pneumatic No. 1; Pneumatic No 2; Pneumatic No 3; Pneumatic 410; Pneumatic Cartridge; Pneumatic Trapshooting Cartridge; Pneumatic Twenty Gauge; Pneumatic Pegamoid.
Example One: Pneumatic No 1. Ga 12; Tc gastight brick red/black; Kn Pneuma Pneumatic Trade Mark. Waterproof. Metal lined. (Bathgate's Patent). Sole manufacturers Pneumatic Cartridge Co Ltd. Telegrams Pneuma Edinburgh; Bl 16; Pr MCI; St ICI; Wd red/black. Pneumatic Cartridge Co · Edinburgh · (Inner) Registered Patent. 4; Or British.
Example Two: Pneumatic Special. Ga 12; Tc greenish grey/black; Kn Pneuma Pneumatic Trade Mark. Loaded with British material throughout. Bathgate's Patent. Trade Marks Regd Nos, 288783 and 290283. Sole manufacturers Pneumatic Cartridge Co Ltd; Bl 8; Pr MCI; St ICI; Wd As on Example one; Or British.

PNEUMATIC CARTRIDGE CO: Bristol, (Avon).
Business: Cartridge experts.
Cartridges: Pneumatic Cartridge.
Example: Pneumatic Cartridge. Ga 12; Tc pink/black; Kn Sole manufacturers, Pneumatic Cartridge Co Ltd, Bristol; Bl 8; Pr MCI; St ICI; Wd brown card/black. (Cluster type) 5; Or British. Note: Some of these cartridges have been seen loaded with Edinburgh top wads and others marked with the shot size on the tube around the top of the base as used for later type star crimp closure.

I have shown the Pneumatic Cartridge Company in three sections. This way I have been able to give examples of cartridges at their appropriate addresses. The Pneumatic Cartridge Co Ltd were in business in Edinburgh from 1904 until 1954, they then made a move south to Bristol and were in business there until 1968. The two patents for their Pneumatic cork wadding were taken out in 1904 and 1910. The centre compression wads were of solid cork being about 11 mm thick and an 8 mm diameter hole running through the centre. This hole was plugged with another solid cork wad about 8 mm thick.

H. E. POLLARD & CO: Broad St, Worcester, Worcs.
Business: Gunmakers.
Cartridges: Gastight; The Keepers Smokeless; The Long Shot; Our Game.
Example: The Keepers Smokeless. Ga 12; Tc light grey/black; Kn Telephone 407; Bl 8; Pr MCI; St ICI; Wd red/black. H. E. Pollard & Cº · Worcester · 5; Or British.

J. E. PONTING: Malmesbury, Wilts.
Business: Ironmonger.
Cartridges: Information given to me in their shop was that many years ago they sold their own cartridges.

WILLIAM POOLE: Market Hill, Haverhill, Suffolk.
Business: Ironmonger.
Cartridges: It is not certain if they sold cartridges with their own name on, but it is certain that they once sold brands of The New Explosives Company around 1914 and prior to the war.

R. & E. POTTER: Thame, Oxon.
Business: Gunsmiths and ironmongers.
Cartridges:
Example: Unnamed. Ga 12; Tc, Grey/black; De A rabbit in full run, FR; Specially loaded R. & E. Potter. Gunsmiths; Bl 8; Pr MCI; St EN; Wd white card/black. Smokeless · Diamond · 6; Or British.

POTTER & CO: 1 Cornmarket, High Wycombe, Bucks.
Cartridges:
Example: Unnamed. Ga 12; Tc orange/black; De A cock pheasant standing, L; Bl 8; Pr MCI; St EL; Or British.

S. E. POTTER & CO: 16, 18, 20 High St, Whitchurch.
Cartridges: The cartridge that I have been told of had the front view of a cat's head on its tube. I have no other details, nor do I know which Whitchurch.

T. POWELL & CO: Salisbury. Wilts.
Cartridges:
Example: Eley's Gas-tight Cartridge Case for Walsrode. Ga 12; Tc purple/yellow; De The EBL shield Trade Mark (With rope-type edging and outlined lettering); Kn Smokeless and waterproof powder; Bl 9; Pr LCI; St T. Powell & Cº. Ltd. Salisbury. Nº.12 Eley; Or British. Though I take this example, which is a fired case, to be from England I have found no reference to T. Powell & Co in Salisbury, Wilts. Therefore it just might be another Salisbury.

WILLIAM POWELL & SON: 35 Carrs Lane, Birmingham 4 (W. Midlands)
Business: Gunmakers.
Cartridges: Admiral; Clay Bird; Ejector; Gastight Metal Lined; General; Knockout; Pegamoid; Specially Hand Loaded; The Steel Lined; Super Velocity.
Example One: Admiral. Ga 16; Tc dark blue/silver; Kn Specially hand loaded by W. Powell & Son. Steel lined. Water resisting. All British; Bl 15; Pr MCI; St ICI; Wd white card/black. Powell · Birmingham · 5; Or British.
Example Two: General. Ga 12; De pink/black; De Powell Trade Mark; Bl 8; Pr MCI; St ICI; Wd white card/black. Powell · Birmingham · 6½; Or British.
Example three: Knockout. Ga 12; Tc eau-de-nil/dark blue; Kn All British smokeless cartridge; Bl 8; Pr MCI; St ICI; Or British.

ALBERT PRATT: Knaresborough, Yorks.
Cartridges: The Fysche.

MARTIN PULVERMANN & CO: 31 Minories, London E.
Business: Wholesale cartridge agents.
Cartridges: Although I have never seen their name on cartridges there is every possibility that they have had their own brand. The MP Ltd. monogram on Mullerite cartridges stands for Martin Pulvermann Limited. (See Muller & Co, also the Mullerite Cartridge Works in this list.) Pulvermanns were the wholesale agents for marketing Clermonite and Mullerite products in the British Isles. Their agency was set up in 1905 and they were in business way into the 1920s and, as such, I have decided to give them a mention.

ARTHUR F. PUNTER: 46 Wote St, Basingstoke, Hants.
Business: Ironmonger.
Cartridges: The Minimax; Shamrock (A Frank Dyke cartridge with extra tube printing bearing Punter's name).
Example: The Minimax. Ga 12; Tc orange/black; Kn Smokeless cartridge. Proprietor, J. M. Emberton; Bl 8; Pr MCI; St ICI; Or British.

Established in 1904, Punters had their name on cartridges until the start of the Second World War. An engraved drawing in one of their old catalogues depicted a cartridge with Punter's name on the headstamping and a cock pheasant on the tube.

JAMES PURDEY & SONS: South Audley St, London, W.1.
Business: Gunmakers.
Cartridges: 2-inch; Eley's Ejector; Pegamoid; Purdey's Deep Shell; Purdey's Large Cap; Purdey's Special.
Example One: Unnamed. Pinfire. Ga 12; Tc olive green/nil; Black powder load; Bl 8; Pr Pinfire; St (Reversed with Ga in centre) Purdey 12; Or Not known.
Example Two: Purdey's Special. Ga 12; Tc brown/black; De Their Trade Mark consisting of four bullets forming a slanting cross; Kn Water resisting. Metal lined; Bl 16; Pr MCI; St EN; Wd yellow/black, E. C. Purdey Loading. 6; Or British.
Example Three: Unnamed. Ga 12; Tc red/black; 50 mm case length; Kn (Tube printing) Purdey London; Bl 8; Pr MCI; St ICI; Wd white card/red. (No writing or shot size but just a red crown. This is a Royal load); Or British.

James Purdey & Sons marketed many cartridges with Bl's 8 and 16 in Ga's 12, 16 and 20. Their Special has been made in all these three gauges. A lot of their cartridges did not carry an individual brand name.

K. D. RADCLIFFE: 150 High St, Colchester, Essex.
Cartridges: A True Fit; Warranted Gastight.
Example: Warranted Gastight. Ga 12; Tc brown/black; De A partridge standing in grass within a circle, FR. Around the top of this is the wording 'Used all the world over'; Bl 16; Pr MCI; St EN: Wd white card/black. (Shot size only in large) 8; Or British.

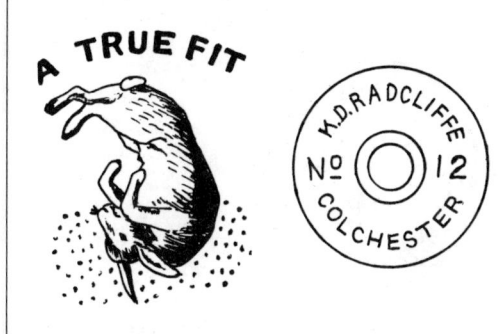

This firm took over the Colchester-based business of J. S. Boreham in November 1899. This means that any cartridge by J. S. Boreham belongs to the previous century.

ROBERT RAINE: Later, RAINE BROTHERS:
Carlisle, Cumberland (Cumbria).
Business: Gunmakers.
Cartridges: The Border Cartridge. Raine's Special.
Example: Raine's Special. Ga 12; Tc brownish orange/black; Kn Loaded by Robert Raine; Bl 8; Pr MCI; St R. Raine. Carlisle. № 12 Eley; Or British.

RAMSBOTTOM: Manchester, Lancs (Gt. Manchester).
Cartridges:
Example: Unnamed. Ga 12; Tc pinkish tan/dark blue; De A large EBL shield Registered Trade Mark (Outlined letters with rope-type edging); Kn Ramsbottom. Manchester. Schultze; Bl 10; Pr; SCI; St Ramsbottom. Manchester. № 12 Eley; Squared turnover; Wd white card/red. Eleys · Loading · 10; Or British.

M. RAY: Dartford, Kent.
Business: Gunmaker.
Cartridges:
Example: Unnamed. Ga 12; Tc pale greyish green/black; Kn Specially loaded by M. Ray. Practical gunmaker; Bl 8; Pr MCI; St KB; Or British.

REDMAYNE & TODD: Nottingham, Notts.
Cartridges: The Champion.
Example: The Champion. Ga 12; Tc red/black; De A standing cock pheasant with the tail in the high position, R; Kn Loaded in Great Britain; Bl 8; Pr L-I; St SS; Or Foreign.

E. M. REILLY & Co: 295 Oxford St, London.
Business: Gunmakers.
Cartridges: The Harewood.
Example One: The Harewood. Ga 12; Tc red/black; Kn Specially loaded by E. M. Reilly & Co. Telegraphic address, Reilly Oxford Street London'; Bl 8; Pr MCI; St EL; Or British.
Example Two: Unnamed. Ga 15; Pinfire; Tc brownish orange/nil; Bl Recorded as Small brass; Pr Pinfire; St (reversed with the Ga in the centre) E. M. Reilly & Co. London. 15; Or Not known.

A few centre-fire cartridges in Ga's 12 and 16 have been seen with their name on the headstamps only. In a list which I bought many years ago, *British Gunmakers Names and Towns 1550 until 1900*, by P. L. Downey it lists Reilly & Co, London 1859. Also, Reilly Edward M., London 1850–1890. All I can say is that any E. M. Reilly & Company cartridge is very old.

REMINGTON (ENGLAND): UMC Works, Brimsdown, Enfield, Middx.
Business: American arms and ammunition manufacturers.
Cartridges: Arrow; Economy; Remington ·410; Remington Kleanbore; Remington Nitro Club.
Example One: Unnamed. Ga 12; Tc dark red/black; Kn loaded at Remington UMC Works Brimsdown England; Bl 8; Pr MCI; St RUR; Or U.S.A.
Example Two: Remington Kleanbore. Ga 12; Corrugated paper tube; Tc summer green/black; Kn British loaded. Case made in U.S.A.; Bl 8; Pr MCI; St RUR; Wd brownish yellow/black. Wetproof smokeless 5; Or, U.S.A. Note: Similar cartridges were also loaded in Ga's 16 and 20.

Kleanbore cartridges were first introduced in America in 1926. Kleanbore was a registered name. James E. Burns, a chemist, invented a non-corrosive primer. He then left his employment with the United States Cartridge Company and sold his idea to Remington. The new primer used lead styphnate, and not only did it not leave harmful residue in the bore, but it also spread a protective coating. Remington held a public contest to choose a name for his new cartridges. Two men from two different parts of the States came up with the name Kleanbore and both were given a prize.

I do not know the extent of the period in which the Remington Arms Union Metallic Cartridge Company (as the firm was sometimes known) were in business in England, but they were active in July 1913.

J. REYNOLDS: Cullompton, Devon.
Cartridges:
Example: Unnamed. Ga 12; Tc brown quality/nil; Bl 8; Pr MCI; St WB; Wd white card/red. J. Reynolds · Cullompton · 6; Or British.

W. RICHARDS: Liverpool & Preston, Lancs.
Business: Gunmaker.
Cartridges: The Castle; The Express; Grand Prix; The Killwell; Kynoch Grouse Ejector; The Mark Down.
Example: Grand Prix. Ga 12; Tc pale orange/black; Kn Telephone No 6925; Bl 8; Pr MCI; St EL; Wd yellow/black. W. Richards · Liverpool · (Inner) Amberite 5; Or British.

RICHARDSON: Dunfermline, Fifeshire
Cartridges: Crieffel.

G. M. RICHARDSON: Dumfries, Dumfriesshire.
Cartridges: Buccleuch; Ideal.
Example: Buccleuch. Ga 12; Tc orange/black; Kn Specially loaded. Phone, 833; Bl 8; Pr MCI; St ICI; Wd brown card/black. (Cluster type) 7; Or British.

WILLIAM G. RICHARDSON: Barnard Castle, Co. Durham.
Business: Cartridge and fishing tackle expert.
Cartridges: The Baliol; The Barnite; The Barnoid.
Example: The Barnoid. Ga 12; Tc blue/black; De A large round crest; Bl 16; Pr MCI; St EN; Or British.

The Barnite. Case colour: The cartridge seen was on an eau-de-nil paper tube with Navy blue print. It is quite possible that it was also produced on a pale green case with black print. Wad colour as illustrated: White card with black print.

JOHN RIGBY & CO: 43 Sackville St, London W.1.
Business: Gun and rifle makers.
Cartridges: Ejector; Rigby's Record Cartridge.
Example One: Unnamed. Ga 12; Tc powder blue/black; Kn John Rigby & Co (Gunmakers) Ltd. 43 Sackville Street. Water resisting; Bl 16; Pr MCI; St EN; Wd white card/black. J. Rigby & Co · London · 6; Or British.
Example Two: Unnamed. Ga 12; Tc pink/black; Kn 32 King Street, St James's, London SW1; Bl 8; Pr MCI; St ICI; Or British.

The firm later made a move to 32 King Street, St James's. The firm was founded in Dublin in 1770.

C. RIGGS & CO: 107 Bishopsgate, London, E.C.
Cartridges: The Bishop.
Example: The Bishop. Ga 12; Tc light brown/black; De A field-type gate with a bishop's mitre on the top bar; Kn Ye Bishops Gate. Sports House, Bl 16; Pr MCI; St EL; Wd white card/black. Smokeless · Diamond · 6; Or British.

A. E. RINGWOOD: Banbury, Oxon.
Business: Gunmaker.
Cartridges: The Dreadnought; The Ideal; The Special.
Example: The Dreadnought. Ga 12; Tc eau-de-nil/dark blue, or greyish green/black; De An oncoming battleship. Bl 8; Pr MCI; St ICI; Wd brown card/black · Smokeless · 6; Or British.

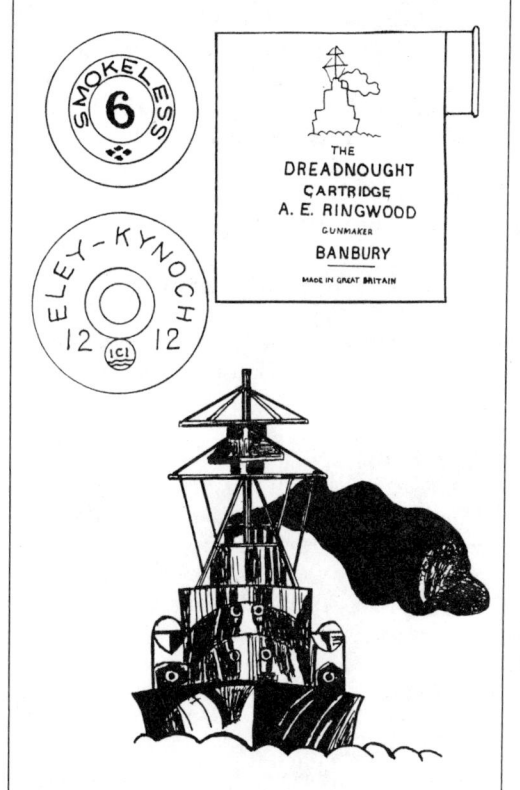

Case colour: Eau-du-nil with Navy blue print or greyish green with black print. Wad colour: White card with red print or brown card with black print.

ROBERTS: Ottery St Mary, Devon.
Business: Ironmonger.
Cartridges: The Ottervale.
Example: The Ottervale. Ga 12; Tc white/black; De A cock pheasant; Bl 8. No other details recorded.

E. ROBERTS: 141 Steelhouse Lane, & 22 Weaman St.
Birmingham (W. Midlands).
Business: Gunmaker.
Cartridges: The Forward Cartridge; The Reliance.
Example: The Reliance. Ga 12; Tc yellow/black; Kn Smokeless cartridge. Practical gunmaker; Bl 8. No other details recorded.

ALEXANDER ROBERTSON & SON: Wick, Caithness.
Business: Ironmongers.
Cartridges: Eley Gas-tight Cartridge Case; Special Smokeless.
Example: Eley Gas-tight Cartridge Case. Ga 12; Tc brown/black; De A large EBL shield Registered Trade Mark (Outlined letters with rope edging); Kn Eley Gas-tight Cartridge Case. Made in Great Britain. Loaded by Alex Robertson & Son. Ironmongers. Wick; Steel lined; Bl 16; Pr MCI; St Robertson & Son.Wick. № 12 Eley; Or British.

ROBINSON BROTHERS: Loftus, Yorks (Cleveland).
Business: Ironmongers.
Cartridges:
Example: Unnamed. Ga 12: Tc red/black; Kn Specially loaded by Robinson Bros. Ironmongers. Loftus; Bl 8; Pr Not recorded; St Robinson Bros № 12 Loftus; Or Not known.

H. ROBINSON: 102 St John's St, Bridlington, Yorks (Humberside).
Cartridges: Burlington Cartridge; Burlington Express.
Example: Burlington Express. Ga 12; Tc middle blue/dark blue; Bl 8; Pr MCI; St ICI; Or British.

H. ROBINSON & CO: Bridgnorth, Salop.
Business: Ironmongers, gunsmiths and cycle makers.
Cartridges: The Castle.
Example: The Castle. Ga 12; Tc middle blue/black; De A pheasant; Bl 8; Pr MCI; St ICI; Or British. No other details recorded.

R. ROBINSON: 7 Queen St, Hull, Yorks (Humberside).
Business: Gunmaker.
Cartridges: Ejector; The Humber; The Kingston Smokeless; The Magnet.
Example: The Kingston Smokeless. Ga 12; Tc grey/dark blue; De An oncoming cock pheasant in flight with his wings in the downward beat, FL; Kn Loaded by R. Robinson (Gunmakers) Ltd; Bl 16; Pr MCI; St ICI; Or British.

R. B. RODDA & CO: Calcutta, India & Birmingham (Midlands).
Cartridges: 3 Inch Long Range; Champion Smokeless; Crown Smokeless; Mullerite Paragon; Rotax Ball Cartridge; The Wellesley.
Example: The Wellesley. Ga 12; Tc orange/black; Bl 8. No other details recorded.

It appears that this firm had more connections with India than it did with Great Britain. Several Rotax Cartridges have come to light in the UK. These have had the word 'Ball' printed on their tubes but have been shot loaded. They may have been First World War loads.

R. ROPER SON & CO: 8 South St, & 9 Exchange St, Sheffield, Yorks.
Business: Gunmakers.
Cartridges: Eley's Gas-tight Cartridge Case; The Hallamshire Cartridge; Kynoch's Patent Perfectly Gas-tight.
Example: The Hallamshire Cartridge. Ga 12; Tc orange tan/black; De Sheffield City's coat of arms; Kn R. Roper Son & Co. Gunmakers. Sheffield. Kynoch's Patent Perfectly Gas-tight; Bl 6 + 17 (Double head); Pr SCI; St KB; Wd red/black. Roper *Sheffield* (Inner) Schultze 5; Or British.

ROSKELLEY: Lostwithiel, Near St Austell, Cornwall.
Business: Ironmonger.
Cartridges:
Example: Unnamed. Ga 12; Tc blue quality/yellow; Kn (Diagnoal printing) Eley London; De An early type EBL shield Registered Trade Mark; Bl 9; Pr SCI; St Roskelley · Lostwithiel · № 12 Eley; Or British.

CHARLES ROSSON: Later, ROSSON & SON:
4 Market Head, & 12 Market Place, Derby.
Business: Gunmakers.
Cartridges: Eclipse; Monvill; Roedich; Vipax.
Example: Monvill. Ga 12; Tc middle blue/black; Kn 4 Market Head, Derby; Bl 16; Pr MCI: St KN; Wd white card/dark blue. Rosson & Son+Derby+ (Inner) Special 4 Loading; Or British.

The firm moved from 4 Market Head to 12 Market Place.

C. S. ROSSON & CO: Rampant Horse St, Norwich, Norfolk.
Business: Gunmakers.
Cartridges: The Crown; The Ektor Long ·410; The Kuvert; The Lowrecoil; The Monvill; The Roedich; The Sixteen Cartridge, Star ·410; The Twenty Cartridge; The Vipax.
Example: Roedich. Ga 12; Tc grey/black; De A setter dog standing, R. (Wording above the dog) 'A good bag assured'; Kn (All the tube printing is around the tube and read when the cartridge is standing on its base except for the name Roedich, which runs diagonally upwards with the word Registered above and Trade Mark below). Loaded only by C. S. Rosson & Co. Gunmakers. Norwich. Tel, Norwich 317. Special Smokeless Cartridge; Bl 8, Pr; MCI; St ICI; Or British.

ROWELL & SON: Chipping Norton, Oxon.
Cartridges: Surekiller.
Example: Surekiller. Ga 12; Tc red/black; Bl 8; Pr -CI; St EL; Or British. No other details recorded.

R. H. ROWLAND: Woodbridge, Suffolk.
Cartridges: Special Loading.
Example: Special Loading. Ga 12; Tc Not recorded; De A cock pheasant in grass, L; Bl Not recorded; St R. H. Rowland · Woodbridge · № 12 Kynoch; Wd Colour not recorded. Smokeless * 5; Or British.

ROWLATTS: 17 Silver St, Wellinborough, Northants.
Business: Gunmaker.
Cartridges: Some notes that I have been given state that this firm was established in 1751 and that in later years it possibly had its name on cartridges. In case this is so I have included it here, although I am unable to find the name in a British gunmakers' list.

ROYS: Wroxham, Norfolk.
Business: Ironmonger.
Cartridges: The De Lux.
Example: Unnamed or name not known. Ga 12; Tc black/silver; De A pheasant; Bl 16. No other details recorded.

A. J. RUDD: Norwich & Great Yarmouth, Norfolk
Business: Gunmaker.
Cartridges: The Norfolk; The Standard; The Star; Rudd's 'X.L.' Cartridge.
Example: Rudd's 'X.L.' Cartridge. Ga 12; Tc orange/black; Bl 8; Pr MCI; St EN; Wd red/black. Rudd's · Loading · 5; Or British.

A. J. RUSSELL: High St, Maidstone, Kent.
Business: Gunmaker.
Cartridges: Russell's Special.
Example: Russell's Special'. Ga 12; Tc brown/black; De A pheasant; Bl 16. No other details recorded.

ALFRED H. RUTT: Cattle Market, Northampton, Northants.
Business: Gunmaker.
Cartridges:
Example: Unnamed. Ga 16; Tc orange/black; Kn Alfred H. Rutt (Late Marsh) Gun Maker; Bl 10; Pr MCI; St A. H. Rutt. Northampton. № 12 Eley; Or British.

R. D. RYDER: Rhayder, Radnor (Powys).
Business: Ironmonger.
Cartridges: Information given to me was that a cartridge was once loaded by them in a bluish-green case.

A. SANDERS: Maidstone, Kent.
Business: Gunmaker.
Cartridges: The Allington; Eley's Gas-tight Cartridge Case; Fourten; Invicta Special; Long Tom; The Medway; Pinfire (R.W.S. case).
Example: Invicta Special. (*above right*). Ga 16; Tc red/black; De A large round crest picturing the Kent prancing horse with the wording A. Sanders Late Swinfen's; Bl 9; Pr MCI; St A. Sanders · Maidstone. № 16 Eley; Or British.

SCHULTZE CO: Also known as,
SCHULTZE GUNPOWDER CO: 28 Gresham St, London.
Business: Gunpowder and cartridge manufacturers.
Cartridges: The Albion; The Bomo; The Captain; The Caro; The Conqueror; Deep Base Gastight; Ejector; The Eyeworth; Grand Prix; Nitro; The Pickaxe; Rainproof; The Torro; Waterproof; The Westminster; The Yeoman.
Example One: The Captain. Ga 12; Tc greyish green/black; Kn The Schultze Co, Ltd; Bl 8; Pr MCI; St EL; Or British. Note: As I once owned in my collection the same cartridge in Ga 20 the chances are that it was also made in Ga 16.
Example Two: Grand Prix. Ga 12; Tc buff/black; De The oval-shaped Schultze Trade Mark with the clenched fist with an electrical flash; Kn The Schultze Gunpowder Co Ltd; Bl 8; Pr MCI; St EL; Wd Colour not recorded. Smokeless; Or British. Note: Example two refers to a Schultze cartridge and not an Eley Brothers, although they manufactured the case.

Schultze, the smokeless gunpowder, acquired

its name from its inventor, Captain E. Schultze. He was an officer in the Prussian Service. The cartridge I have given as Example one, called The Captain, honours this famous gentleman. The Schultze Gunpowder Company Limited was formed in Britain in 1868 and they had a gunpowder mill at Eyeworth Lodge in the New Forest area. The cartridge called The Eyeworth was named after this factory.

I am very sceptical about the date when The Schultze Gunpowder Co finally merged with Messrs Eley Brothers. I believe it started around 1911 and that they still produced their brands of cartridges under their own name until 1923. I have found a note in my files stating that Schultze's own cartridge dated from 1899 until 1923, when the company was finally liquidated. If this is so, then it brings this period within the control of Nobel Industries Limited. It is all very confusing. Also, according to my notes, at the time when the company merged with Eley's during 1911 it changed its name to The Schultze Company Limited. It also states that later on in 1916, when the Great War was getting bloody, a very strong anti-German feeling running through the country made them decide to revert back to being known as The Schultze Gunpowder Company Limited. If this is so, then the cartridges, such as Example one, can be dated as having been made between 1911 and 1916.

In 1909 a separate company was formed. To give them their full title, they were Cogschultze Ammunition & Powder Company Limited. Their name can be found under this heading in this list. An arrangement was made whereby Cogswell & Harrison provided the cartridge cases and Schultze the powders. In an article I once read it said that this concern probably went on until the start of the First World War, but I believe it only ran for two years. As Eley Brothers had supplied Schultze with cartridge cases until 1909, and as it had been reported that they merged with them in 1911, this could very likely have been the reason that brought it all about.

When firms joined the combine of Messrs Eley Brothers, it was common practice often to retain some of the brand and trade names of their cartridges and components. Two of the brand names of Schultze, Westminster and Yeoman, continued to be produced within the combine right up to 1939.

Other powders marketed by The Schultze Gunpowder Company, apart from the original Schultze, are listed here with their year of introduction.

Powders: Imperial Schultze, 1902; Cube Schultze, 1908; Popular, 1912 approx; Lightning, 1913.

This information can help in dating old cartridges. For instance, the Popular Powder dated from 1912, the merger with Messrs Eley Brothers being 1911, then the Grand Prix, a name of Eley origin, places this cartridge somewhere between 1912 and 1923.

An old advertisement for the wholesale of Schultze's powders gave an address at 3 Bucklersbury, London, E.C. This was the Company's offices.

JOHN A. SCOTCHER & SON: 4 The Traverse, Bury St Edmonds, Suffolk.
Cartridges: Eley's Gas-tight Cartridge Case for Schultze Gunpowder; The Invincible.

This business was taken over by Henry Hodgson in 1913.

SCOTT & SARGEANT: 26 East St, Horsham, Sussex.
Business: Ironmongers.
Cartridges: The Horsham Special; The Ironmonger.
Example: The Horsham Special. Ga 12; Tc orange/black; De Hen pheasant squatting with head and tail raised, L; Kn British-loaded smokeless; Bl 8; Pr LCI; St SS; Wd Colour not recorded. Special · Smokeless · 6; Or Foreign.

F. A. SHARP & SON: Poole, Dorset.
Business: Ironmongers.
Cartridges: Sharp's Express.

J. S. SHARPE: 35 Belmont St, Aberdeen.
Business: Gunmaker.
Cartridges: The Scottie.
Example: The Scottie. Ga 12; Tc dark green/black; De Trade Mark was a black Scottish Terrier; Kn Smokeless cartridge. Specially loaded by J. S. Sharpe. Gun and fishing tackle maker. Telephone 4066; Bl 8; Pr MCI; St ICI; Wd red/black. Special · Smokeless · 6; Or British.

SHUFFREYS: Walsall, Staffs (W. Midlands).
Cartridges: The Beacon.
Example: The Beacon. Ga 12; Tc brown/black; De A running rabbit bounding over a small foxglove plant, Fr; Kn Specially loaded; Bl 9; Pr MCI; St KB; Wd brown card/nil. Loaded with shot; Or British.

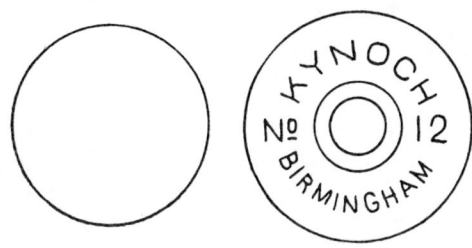

Case colour: Light brown quality paper with black print. Wad colour as on the example: Plain pale brown card with no printing. The example was loaded with shot.

S. W. SILVER & CO: London.
Cartridges: A cartridge bottom found in India was as follows: Pr MCI; St S. W. Silver & Co. Nº 12 London.

SIMPSON: Piccadilly, London.
Cartridges:
Example: Unnamed or name not known. Ga 12; Tc yellow/black; Bl 16. No other details recorded.

SKINNER & CO: 63 Haywood St, Leek, Staffs.
Cartridges:
Example: Unnamed, Ga 12; Tc brown quality/black; Bl 7; Pr SCI; St KB; Wd white card/black. Skinner & Cº · Leek · 6; Or British.

SLATER: Warwick St, Leamington Spa, Warwickshire.
Cartridges: I have been given their name for this list but have no other information.

SLINGSBY GUNS: Boston & Sleaford, Lincs.
Cartridges: Slingsby's Champion; Slingsby's Fen; Slingsby's Special; Slingsby's Stump.
Example: Slingsby's Champion. Ga 12; dark green/black or dark blue; De A cock pheasant standing with his tail horizontal, L; Kn Long-distance cartridge loaded with high-grade powder; Bl 16; Pr MCI; St ICI; Wd yellow/black. Smokeless · Diamond · 6; Or British.

JOHN SMAIL & SONS: Morpeth, Northumberland.
Business: Ironmongers.
Cartridges: The Lightning Killer.
Example: The Lightning Killer. Ga 12; Tc middle blue/dark blue; De A flying grouse; Kn Best smokeless; Bl 8; Pr MCI; St ICI; Or British.

SMALLWOOD: 12 High St, Shrewbury, Salop.
Cartridges: Smallwood's Challenge.
Example: Smallwood's Challenge. Ga 12; Tc Red/black; Kn Smokeless. Phone 3081. Steel lined. Special loading; Bl 8; Pr LCI; St Jas R. Watson & Cº 12 12 London; Wd yellow/black. Special · Smokeless · 4; Or Belgium.

A. F. SMITH: Hailsham, Sussex.
Cartridges: The Hailsham Special.
Example: The Hailsham Special Ga 12; Tc red/black; De A cock pheasant standing in grass with the tail horizontal, R; Kn Smokeless cartridge; Bl 8; Pr LCI; St SS; Wd red/black. Special · Smokeless · 5; Or Foreign.

C. H. SMITH & SONS: 123 Steelhouse Lane, Birmingham, (W. Midlands).
Cartridges: The Abbey; The Invincible.
Example: The Abbey. Ga 12; Tc orange/black; De A cock pheasant: Bl 8; Pr MCI; St EN or ICI; Or British.

CHAS SMITH & SONS: 47 Market Place, Newark, Notts.
Business: Gunmakers.
Cartridges: The Castle; The Clinton; All British Extra Special; The Newark Cartridge; The Rufford; Schultze Loaded; The Universal.
Example: The Castle. Ga 12; Tc brownish orange/black; Kn Telephone Newark 228. Foreign made case. Loaded in England; Bl 8; Pr LBN; St SS; Wd green/black. Smith & Sons · Newark · IX; Or Foreign.

Case colour: Brownish orange with black print. Wad colour as illustrated: Summer green with black print.

STEVE SMITH: 42 High Friar St, Newcastle-under-Lyme, Staffs.
Cartridges: Trap and Game.
Example: Trap and Game. Ga 12; Tc orange/black; De A flying clay and standing setter dog, L; Bl 9; Pr MCI; St MGB; Wd green/black. Special · Smokeless · 6; Or British.

SMITH MIDGLEY: Bradford, Yorks.
Cartridges: Pegamoid.
Example: Pegamoid. Ga 12; Tc light brown/black; Bl 16; Pr MCI; St Eley N.I. № 12 London; Wd Name of the firm on the top wad only; Or British.

SMOKELESS POWDER & AMMUNITION CO: address unknown.
Business: Gunpowder and ammunition manufacturers.
Cartridges: Ejector.
Example: Unnamed. Ga 12; Tc maroon/nil; Bl 9; Case length 70 mm; Pr SCI; St S. P. & A. № 12 Cº; Or Not known.

A note in my files states that this company was formed in 1898.

J. F. SMYTHE: Darlington & Stockton-on-Tees (Cleveland).
Business: Gunmaker.
Cartridges: Smythe's Champion; Durham Ranger; Ejector; The Field; Gastight; Smythe's Special Load.
Example: Smythe's Special Load. Ga 12; Tc purple/silver; Bl 8; St Smythe 12 Darlington, Stockton-on-Tees. No other details recorded.

H. & R. SNEEZUM: 14, 16, 18, 20 Fore St, Ipswich, Suffolk.
Business: Gunmakers.
Cartridges: Sneezum's Anglia; Sneezum's Special High Velocity Load.
Example: Sneezum's Anglia. Ga 12; Tc orange/black; De A running hare; Kn Telephone Ipswich 3923; Bl 8; Pr MCI; St ICI; Or British.

Enlarged drawing from the Special

Enlarged drawing of game on the Anglia

SOUTHERN COUNTIES AGRICULTURAL TRADING SOC: Winchester, Hants.
Business: Agricultural Traders.
Cartridges: The Challenger Smokeless.
Example: The Challenger Smokeless. Ga 12; Tc dull burgundy/silver; De A hen pheasant, L; Bl 8; St SS; Or Foreign. No other details recorded.

This company's name is often abbreviated as S.C.A.T.S. and is called Scats.

J. W. & E. SOWMAN: Olney, Bucks.
Business: Ironmongers.
Cartridges: The Sureshot Smokeless.
Example: The Sureshot Smokeless. Ga 12; Tc grey/black; Kn Special loadings; Bl 8; Pr MCI; St KN; Wd orange/black. E.C. · Powder · 5; Or British.

Case colour: Grey with black print. Wad colour as illustrated: Deep orange with black print.

ALFRED L. SPENCER: Richmond, Yorks.
Business: Gunmaker.
Cartridges:
Example: Unnamed. Ga 12; Tc dark green/dark blue; De A standing partridge, L; Kn Hand loaded and guaranteed by Alfred L. Spencer. Gunmaker; Bl 16; Pr MCI; St ICI; Wd white card/black. (The printing forming a background for the white wording) A. L. Spencer · Richmond · 6; Or British.

A top wad has been seen loaded into a case with the St ICI and the name A. A. Spencer · Richmond.

F. P. SPENCER: Lugley St, Newport, Isle of Wight.
Cartridges: F.P.S. Vectis Special Loading; Spencer's Vectis Bunnie; Spencer's Vectis Special.
Example: Spencer's Vectis Special. Ga 12; Tc dark green/black; De The Isle of Wight shield (coat of arms); Bl 16. No other details recorded.

SPORTING PARK: London.
Cartridges: Eley Ejector.
Example: Eley Ejector. Ga 12; One-piece brass case with a brown inner paper tube; Bl 57; Pr MCI; St Eley №12 Ejector; Wd turquoise/Black. London. Sporting Park. (Shot size in purple ink) 8; Or British.

S. SPORTRIDGE (G.B.): Address unknown
Cartridges: The Sportridge.
Example: The Sportridge. Ga 12; Tc dark red/black; Bl 6; Pr MCI; St 12 12; Wd light blue/dark blue. S. Sportridge. (G.B.) Ltd · 5; Or Not known.

STACEY: Dulverton, Somerset.
Cartridges: Cartridge remains have been found with St Stacey №12 Dulverton.

STANBURY & STEVENS: Alphington St, Exeter, Devon.
Cartridges: The Devonia; The Game; The Monocle; The Red Flash; The Stanby; The Swift.
Example: The Swift. Ga 12; Tc greyish green/black; Bl 8; Pr MCI; St ICI; Or British.

STEBBINGS: Attelborough, Norfolk.
Cartridges: I have seen the following top wad. Wd orange/black. Stebbings · Attleboro · 4.

T. STENSBY & CO: 6 Withy Grove. Later 12 Withy Grove, Manchester 4, Lancs (Gt. Manchester).
Business: Gun, rifle and cartridge makers.
Cartridges: The All British; The Champion Gastight.
Example One: Unnamed. Ga 16; Tc brown/black; Kn 6 Withy Grove. Gun, rifle and cartridge makers. Telephone No 6589. Telegrams, Stensby Manchester; Bl 16; Pr MCI; St ELG; Or British.
Example Two: The All British. Ga 12; Tc gastight brick red/black; Kn Gastight and metal lined. Water resisting. Loaded by T. Stensby & Co. 12 Withy Grove. Telephone No 6589 Blackfriars. Telegrams, Stensby, Gunmaker Manchester; Bl 16; Pr MCI; St ICI; Wd white card/red. Smokeless · SSG; Or British.

STERLING: London.
Cartridges: Sterling.
Example: Sterling. Ga 20; Tc middle blue/dark blue; De A crown Trade Mark; Kn Extra prima. Smokeless cartridge; Bl 15; Pr LCI; St Sterling 20 20 London; Wd white card/black. Special Smokeless · 5; Or British.

STILES BROTHERS: Warminster, Wilts.
Cartridges: Kill Quick.
Example: Kill Quick. Ga 12; Tc orange/black; De A cock pheasant standing in grass with tail horizontal, L; Kn Smokeless; Bl 8; Pr MCI; St ICI; Wd yellow/black. Smokeless · Diamond · 6; Or British.

A. J. STOCKER & SON: Also **C. & E. STOCKER**: Chulmleigh, Devon.
Business: Ironmongers.
Cartridges: The Chulmleigh.
Example: The Chulmleigh. Ga 12; Tc pale apple green/black; De A cock pheasant standing on a tuffet, L; Kn Bracketed (C & E Stocker); Bl 8; Pr MCI; St ICI; Or British.

F. E. STOCKER: St Austell, Cornwall.
Cartridges: I regret that I have no cartridge information on this firm.

A. STOKES: Hastings, Sussex.
Cartridges: This name has been seen on a top wad that was loaded into a Nobel's Sporting Ballistite case.

E. R. STRICKLAND & SON: Gillingham. (County unknown.)
Cartridges: The Gillingham Cartridge; The Quick Fire.
Example: The Gillingham Cartridge. Ga 12; Tc red/black; De A running rabbit; Bl 8; Pr MCI; St KB; Or British.

J. STRONG & SON: 65 Castle St, Also 8 Warwick Rd, Carlisle, Cumberland (Cumbria).
Business: Gun and ammunition dealers.
Cartridges:
Example: Unnamed. Ga 12; Tc orange/black; De A picture of half a dozen pheasants feeding in a woodland ride; Kn Telephone № 796; Bl 8; Pr MCI; St ICI; Wd yellow/black. Smokeless · Diamond · 6; Or British.

STUCHBERYS STORES: Maidenhead, Berks.
Business: Stores.
Cartridges:
Example: Unnamed. Ga 12; Tc dull orange/black; De, A large and long letter 'S' used to start the first three words on the tube printing; Kn, Stuchbery's Stores Special. Maidenhead; Bl 8; Pr MCI; St EL; Or British. Note: The example seen was a window cartridge display round. A window let into the tube showed the internal components.

SWINFEN: Maidstone, Kent.
Business: Gunmaker.
Cartridges: Swinfen's Special.
Example: Swinfen's Special. Ga 20; Tc dull red/black; Bl 10; Pr MCI; St ELG; Or British. This example was an unused ready-capped case.
The firm of Swinfen was later taken over by A. Sanders.

SYKES BROTHERS: Ossett, Yorks.
Cartridges: Eley Ejector.
Example: Eley Ejector. Ga 12; One-piece brass case with a crimson inner paper tube; Bl 57; Pr MCI; St Eley № 12 Ejector; Wd white card/red. Sykes Bros · Ossett · 6; Or British.

TAYLOR: Driffield, Yorks.
Cartridges: I have been told that a fired cartridge case has been seen with this name printed on its tube.

A. TAYLOR: 49 Bartholomew St, Newbury, Berks.
Business: Barber and field sports.
Cartridges:
Example: Unnamed. Ga 12; Tc yellow/black; Kn A. Taylor. Mullerite Smokeless. Newbury; Bl 8; Pr LCI; St SS; Or Not known.

J. T. TAYLOR: Bromsgrove, Worcs (Heref & Worcs).
Business: Ironmonger.
Cartridges:
Example one: Unnamed. Ga 12; Tc ruby red/black; Bl 8; Pr MCI; St RUR; Or U.S.A.
Example two: Unnamed. Ga 12; Tc dark red/black; Kn J. T. Taylor & Son. Ironmongers. Bromsgrove; Bl 10; Pr MCI; St RUR; Wd brown card/black. Remington. Wetproof. 5; Or U.S.A. Note: Though similar to that already listed, the name Son is now shown in the business.

H. G. TETT: Coventry, Warwickshire (W. Midlands).
Business: Gunsmith.
Cartridges:
Example: Unnamed. Ga 12; Tc light blue/black; Kn Loaded with Curtis's & Harvey's Smokeless Diamond. Phone Coventry 73; Bl 16; Pr MCI; St EN; Or British.

THACKER & CO: Worcester, Worcs.
Business: Gunmakers and fishing tackle dealers.
Cartridges: Eley Gastight Quality; Long Shot Smokeless.
Example: Long Shot Smokeless. Ga 12; Tc cream white/black; De A rabbit in full run bounding over a small foxglove plant, FL; Kn Telephone № 407; Bl 8; Pr MCI; St KN; Wd white card/red. Special · Smokeless · 5; Or British.

THOMPSON BROTHERS: Bridgwater, Somerset.
Business: Ironmongers.
Cartridges: Ruby.
Example: Ruby. Ga 12; Tc brownish orange/black; Kn Special smokeless; Bl Not recorded; Pr -CI; St EL; Or British.

W. J. TICKNER: Bishops Waltham, Hants.
Cartridges: The Sportsman.
Example: The Sportsman. Ga 12; Tc ruby red/black; Kn The name of the cartridge diagonally printed on the tube. Loaded for W. J. Tickner; Bl 7; Pr -CI; St SS; Or Foreign.

S. TILBURY & F. A. JEFFRIES: Parson's Garage, Littlehampton Rd, Worthing, Sussex.
Business: Believed to have been garage proprietors.
Cartridges: The Highdown.
Example: The Highdown. Ga 12; Tc purple paper tube, print colour not recorded. De Standing hen pheasant with head and tail raised, L; Kn Rust resistant. Special smokeless. British hand loaded. Phone Swandean, 35; Bl 8; Pr MCI; St SS; Wd Colour not recorded. Special · Smokeless · 5; Or Foreign.

WILLIAM C. TILL: Battle, Sussex.
Business: Ironmonger.
Cartridges:
Example: Unnamed. Ga 12; Tc orange/black; Bl 8; Pr MCI; St KB; Or British.

R. TILNEY & SON: Beccles, Suffolk.
Business: Gunsmiths.
Cartridges: Tilney's Special.
Example: Unnamed. Ga 12; Tc salmon pink/black; Kn R. Tilney & Son. Gunmakers. Beccles; Bl 8; Pr MCI; St Nobel 12 12 Glasgow; Wd Colour not recorded. Tilney & Son · Beccles · 5½; Or Scottish. Note: This cartridge tube carries the word Gunmakers.

TILY & BROWN: Guildford & Farnham, Surrey. Also at Farnborough, Hants.
Cartridges: Farnford.
Example: Farnford. Ga 12; Tc dark red/black; Bl 8; Pr MCI; St RUR; Wd light brown card/black. Remington Smokeless Powder (Inner) Wetproof 6; Or U.S.A.

F. H. TIMS: Truro, Cornwall.
Cartridges:
Example: Unnamed. Ga 12; Tc orange/nil; Bl 8; St F.H.Tims Truro. No other details recorded.

JOHN TINNING: Longtown, Cumberland (Cumbria). Also at Newcastleton, Roxburghshire.
Cartridges:
Example: Unnamed. Ga 12; Tc pale yellow/dark blue; Kn John Tinning. Longtown, and Newcastleton N.B.; Bl 9; Pr MCI; St KB; Or British.

TRENT GUN & CARTRIDGE WORKS: Grimsby, Lincs (Humberside).
Business: Gun and cartridge manufacturers.
Cartridges: A.E.C. Rook; Best Smokeless; Deep Shell; Favourite; London; Super Range.
Example: Best Smokeless. Ga 16; Tc ruby red/black; De The action of a shotgun showing an ejecting case, L; Kn Foreign made. Loaded in England with British shot; Bl 6; Pr LCI; St SS; Wd red/black. British · Shot · 5; Or Foreign.

Known to many as Trent's Shot Tower, Grimsby. The Trent Gun and Cartridge Works also produced brands for a few smaller firms. Their large cartridge-loading factory was built in 1929. Back in the 1930s very many firms stocked Trent's cartridges including some small village stores. These cartridges were cheaper than many others for their standard and were very much liked and sold well. The firm went into liquidation in 1953 and closed its door in April. Sad to say in the latter days its loadings were not quite so consistent. At the end batches of their cartridges were sold off in lots. It was then found that a few of these contained only shot but, worse still, a few were filled with nearly all powder. These cartridges were then quickly called back. I remember when I was a young lad that my father used and liked them. I also remember during the first years of the Second World War going beating. Cartridges were then in short supply and a walking gun next to me was using Trent's Deep Shell. Every time he fired his gun there was a colossal bang with a large sheet of red flame, and a pall of black smoke drifted through the trees. His empty cartridge cases were found to be charred all around their tops.

S. TROUGHTON: 24 Caunce St, Blackpool, Lancs.
Business: Gunmaker.
Cartridges:
Example: Unnamed. Ga 12; Tc red/black; De A pheasant; Kn Practical gunmaker; Bl 8; Pr MCI; St KB; Wd orange/black. S. Troughton · Blackpool · 6; Or British.

TRULOCH & HARRISS: Dublin, Republic of Ireland.
Cartridges:
Example: Unnamed. Ga 12; Tc brown/black; Kn Telegram, Shooting Dublin; Bl 16; Pr -CI; St Truloch & Harriss. Dublin. № 12; Wd Reported to have had their name on; Or Not known.

TURNBULL: Bridgnorth, Salop.
Cartridges:
Example: Unnamed. Ga 12; Tc salmon pink/black; Kn Turnbull. Bridgnorth. (No other tube printing); Bl 8; Pr Not recorded; St Turnbull Bridgnorth № 12 Joyce; Wd Colour and shot size not recorded. Turnbull · Bridgnorth; Or British.

ARTHUR TURNER: 5 West Bar, Sheffield, Yorks.
Business: Gunmaker.
Cartridges: The Alliance; The Clay Bird; The Double Wing; The Steeltown; The Wing; The Wizard.
Example: The Wizard. Ga 12; Tc spring green/black; Kn The Wizard. Registered Trade Mark. Loaded by Arthur Turner. Late Maleham & Co; Bl 8; Pr MCI; St ICI; Wd light green/black. Turner · Sheffield · 5; Or British.
The name of this firm was C. H. Maleham & Co until 1920, when it was changed to Arthur Turner.

HENRY A. TURNER: Marlborough, Wilts.
Business: Gunmaker.
Cartridges: The Kennett
Example: Unnamed. Ga 12; Tc brownish orange/black; De A cock pheasant standing in grass, L; Kn Henry Turner. Gunmaker. Marlborough; Bl 8; Pr MCI; St * Eley's – Grand Prix * 12 (Similar to EGP); Wd orange/black. E.C. 6; Or British.

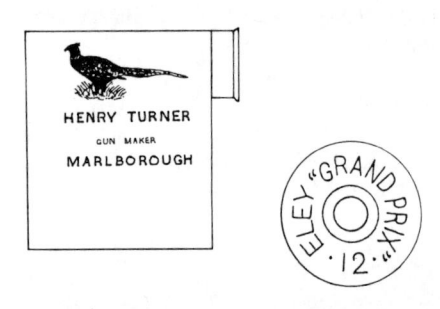

HENRY TURNER
GUN MAKER
MARLBOROUGH

ELEY "GRAND PRIX" ·12·

Case colour: Light orange with black print.

J. TURNER: Penrith, Cumberland (Cumbria).
Cartridges:
Example: Unnamed. Ga 12; Tc dark blue quality/white; Kn Eley London; Bl 10; Pr SCI; St J. Turner № 12 Penrith; Wd white card/red. (The printing forming a background for the white wording) J. Turner · Penrith · ; Or British.

THOMAS TURNER & SONS: 8 Butter Market, Reading, & 16 Northbrook St, Newbury, Berks. Also at 35 Wote St, Basingstoke, Hants. Now at 208 Gosbrook Rd, Caversham, Reading.
Business: Gun and legging makers. Also cartridge loaders.
Cartridges: British Wonders; The Craven; Ejector; The Fillbag; The Grey Rapid; Midget ·410; The Penwood; The Renowned; Special Loading; Turner's Smokeless Wonder.
Example One: Unnamed. Pinfire. Ga 12; Tc dark blue quality/nil; Bl 11; Pr Pinfire; St Turner & Sons . Reading . · No12 · ; Wd white card/red. (The printing forming a background for the white wording) Turner · Reading · 5; Or British.
Example Two: Special Loading. Ga 12; Tc light yellow/black; Kn T. Turner & Sons. Gunmakers. Reading & Newbury; Bl 10; Pr SCI; St Turner. Reading & Newbury. № 12 Eley; Wd white card/red. (The printing forming a background for the white wording) Turner · Reading · ; Or British.
Example Three: Unnamed. Ga 20; Tc light yellow/black; De The then unofficial Borough of Newbury coat of arms. This was the gate house from the long extinct Newbury Castle; Kn T. Turner & Sons. Gunmakers. Newbury. Special smokeless cartridge; Bl 7; Pr SCI; St (Outer) Turner. Reading & Newbury. (Inner) Eley № 20; Or British.
Example Four: The Fillbag. Ga 12; Tc orange/black; De The oval-shaped Common Seal of Reading. This had five male heads; Kn T. Turner & Sons Ltd. Gunmakers. Reading, Newbury and Basingstoke; Bl 8; Pr MCI; St ICI; Wd yellow/black. Turner · Reading, Newbury & Basingstoke · 5; Or British.

Thos Turner & Sons Ltd at one time loaded all their own brand cartridges plus some without brand names. Later, like most firms, their cartridges were loaded for them by the large manufacturers. Turners first started business in Reading and then extended to Newbury and still later to Basingstoke. In their later years they sold off the Reading shop. The new owners bought the goodwill and traded on in the name of Thomas Turner & Sons (Reading) Ltd. The Newbury and then the Basingstoke branches were sold to J. C. Cording & Co, who moved west from Piccadilly, London.

Over the years, some of the branded cartridges became associated with the various branches. Reading had the Wonder and the Renowned; they may also have had the ·410 Midget, made in long and short. Newbury had the Ga 20 that I have listed in Example three. They also had a cartridge called the Penwood, taking its name from a wood to the south of Newbury where the firm set up a shooting school. I have never seen one of these; they were only sold at the school and this was short-lived due to the the Second World War. One brand that was sold by the Basingstoke branch was the Grey Rapid.

Turner's cartridges throughout the many years often displayed on their tubes the Common Seal of Reading. This old seal depicted five male heads in a semi-oval frame. The Reading coat of arms is very similar, but this portrays five maidens' heads. The Newbury shop in its earlier days had at least one cartridge (See Example three) with the Newbury castle gate house. In later years Turner's re-used this gate house by placing it in a circle and portraying it on the last of their Fillbags; this was in post-Second World War years. They also used it on a cartridge called the Turnax which had a star crimp closure. It was Turner's own answer to the Eley Impax. Their shop in Basingstoke also marketed a Turnax at this time, but it had a pheasant on its tube.

All the cartridges marketed by J. C. Cording & Co were of the star crimp closure type, hence their name is not included in this cartridge list. They first started off with a cartridge called the Beater. Later they stopped this and reintroduced the old Turner names Fillbag and Turnax.

An old advertisement from the Newbury Weekly News, No. 2,240 and dated Thursday March 3, 1910, lists the following: Nitrones at 9/- per 100; Bonax at 7/10 per 100; Fillbags at 7/6 per 100; Cravens at 10/- per 100. The Craven took its name from a very large country estate to the west of Newbury and also from the famous local hunt.

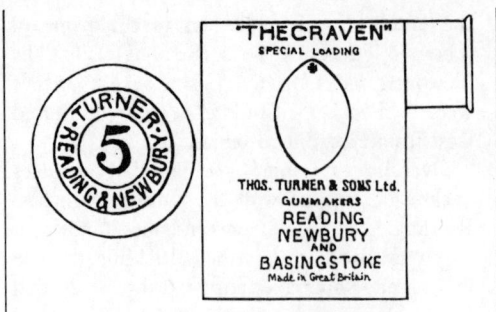

The Craven. Case colours: Dark or forest green with black or Navy blue print. Wad colours: White card or light yellow with black print.

TURNERS CARBIDES: 58 De Grey St, Hull, Yorks (Humberside).
Business: Cartridge loaders and marketers.
Cartridges: The ·410; The Killer; The Standard; The Super: The Standard and The Super were known to have been loaded in Ga's 12, 16 and 20. The Killer, also known as the Keepers Cartridge, was only loaded in Ga 12. An old advertisement lists Standards at 14/- per 100 and Supers at 16/- per 100. Unfortunately the date of this old advert is missing. The Hull Cartridge Company now reside at the above address.

J. A. Twyble: Portadown, County Down, Northern Ireland.
Cartridges: The Invincible Long Range.

JOHN TYLER: Highbridge, Somerset.
Business: Gun and ammunition specialist.
Cartridges: The Falcon.
Example: The Falcon. Ga 12; Tc crimson/black; De A covey of eight partridges flying but showing no ground, FL; Kn John Tyler (Highbridge) Ltd. Gun and ammunition specialists. Phone, Highbridge 18; Bl 16; Pr MCI; St ICI; Wd bright yellow/black. Tyler · Highbridge · 6; Or British.

UNDERHILL: Newport & Eccleshall, Staffs.
Cartridges:
Example: Unnamed. Pinfire. Ga 12; Tc brown/nil; Bl 9; Pr Pinfire; St Underhill. Newport & Eccleshall. № 12; Wd Colour not recorded. K. S. Powder. 5; Or Not known.

W. URTON: Chesterfield, Derbyshire.
Cartridges:
Example: Unnamed (Possibly known as The Spire). Ga 12; Tc brown/black; De The twisted spire; Kn Water resisting; Bl 16; Pr MCI; St EN; Wd white card/black. Smokeless · Diamond · 5; Or British.

J. C. VAUX: Hanwell, Near Banbury, Oxon.
Cartridges:
Example: Unnamed. Ga 12; Tc light green/nil; Bl 5; Pr SCI; St № 12; Wd red/black. J. C. Vaux · Hanwell · 8; Or Not known.

J. VENABLES & SON: Oxford, Oxon.
Business: Gun and rifle makers.
Cartridges: ·410 Long and short; 12 Ga pinfire Unnamed; The County; The Oxford; The Sixteen.
Example: The Oxford. Ga 12; Tc middle blue/dark blue or black; De The Oxford City coat of arms; Kn Loaded only by J. Venables & Son. Gun and rifle makers. Telephone No. 4257; Bl 16; Pr MCI; St ICI; Wd orange/black. Schultze • • BB; Or British.

H. J. WADDON: Wedmore, Somerset.
Cartridges: Special Smokeless Cartridge.
Example: Special Smokeless Cartridge. Ga 12; Tc yellow/black; De A dark walking cock pheasant with a high tail, R; Bl 14; Pr; LCI; St Extra Deep 12 12 Gastight; Or Not known. Note: This example was taken from an unused case.

JOHN WADDON & SONS: Bridgwater, Somerset.
Cartridges: The Quantock.
Example: The Quantock. Ga 12; Tc yellow/black; De A cock pheasant standing on a clump of grass with the tail horizontal, L; Bl 8; St and Wd Not recorded; Or British.

WADDON & SONS: Bridgnorth, Salop.
Business: Ironmongers.
Cartridges:
Example: Unnamed. Ga 12; Tc white/black; De A cock pheasant standing; Bl 8; Pr MCI; St KN; Wd Colour not recorded. Smokeless · Diamond; Or British.

DERRIAN WALES: Great Yarmouth, Norfolk.
Business: Ironmonger.
Cartridges:
Example: Unnamed. Ga 12; Tc blue/black; Bl 10; Pr SCI; St D.Wales. Yarmouth. Nº 12 Eley; Or British.

H. WALKINGTON: Bridlington, Yorks (Humberside).
Business: Dealer in guns and ammunition.
Cartridges: The Reliable.
Example: The Reliable. Ga 12; Tc ruby red/black; Bl 8; Pr MCI; St RNC; Or U.S.A.

D. H. WALLAS: Wigton & Carlisle, Cumberland (Cumbria).
Business: Gunmaker and cartridge loader.
Cartridges:
Example: Unnamed. Ga 12; Tc brownish orange/black; De A cock pheasant with his tail horizontal, L; Kn Best smokeless; Bl 8; Pr MCI; St KB; Wd red/black. D. H. Wallas · Wigton · 6; Or British.

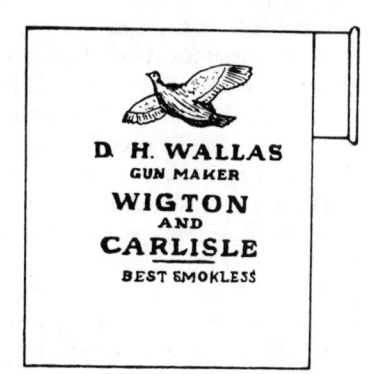

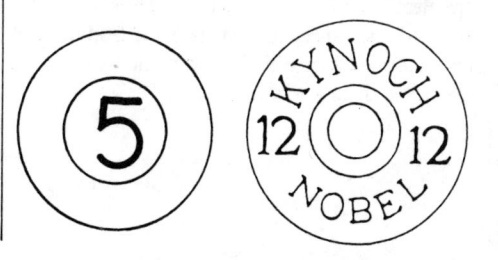

(*Below left*): Case colour: Dull orange with black print. Wad colour as illustrated: White card with black print.

W. WALLAS: Wigton, Cumberland (Cumbria).
Business: Gunmaker.
Cartridges:
Example: Unnamed. Ga 12; Tc turquoise/nil; Bl 8; Pr MCI; St Nº 12; Wd white card/red. (The printing forming the background for the white wording) W. Wallas · Wigton · 5; Or Not known.

WALLIS BROTHERS: Lincoln, Lincs.
Business: Gunmakers.
Cartridges:
Example: Unnamed. Ga 20; Tc buff/black; De The Lincoln City coat of arms; Kn Specially loaded by Wallis Brothers. Gun makers, Lincoln. Gastight cartridge case; Bl (Double head) 9 + 17; Pr MCI; St (EBL Shield) Nº 20 Eley London; Wd white/card/black. (Cluster type) 6; Or British.

WALLIS BROS & SKEMPTONS: Lincoln, Lincs.
Business: Gunmakers.
Cartridges: The Big Tom.
Example: The Big Tom. Ga 12; Tc orange/black; Kn Specially loaded. Tel Lincoln 502; Bl 8; St Not recorded; Wd deep green/black. Wallis Bros · Lincoln · 6; Or Not known.

W. WANLESS: 20 Norfolk St, Sunderland, Co. Durham (Tyne & Wear).
Cartridges:
Example: Unnamed. Ga 12; Tc brown/black; De A coat of arms; Kn W. Wanless Gunmakers; Bl 8; Pr -CI; St Wanless · Sunderland. Nº 12 Eley; Wd Colour not recorded. Wanless · Sunderland · 6; Or British.

Note: I have been given two addresses for this firm. They are 20 Norfolk Street and 29 Norfolk Street. It is possible that one of these may be wrong.

WANLESS BROTHERS: Sunderland, Stockton-on-Tees & South Shields (Tyne & Wear.) Also Cleveland.
Business: Gun and rifle makers.
Cartridges: The Long Range; The Waterloo.
Example: The Waterloo. Ga 12; Tc light brown/black; De The names, Sunderland and that of Stockton-on-Tees both start by using the same large letter 'S'. The name Waterloo is in longhand; Bl 16; Pr MCI; St KB; Wd orange/black. Wanless. Sunderland & Stockton. 5; Or British.

This firm loaded several very decorative cartridges. Often they had no brand names and were crested.

WARD & SON: Worcester, Worcs.
Cartridges: Kynoch Grouse Ejector.
Example: Kynoch Grouse Ejector. Ga 12; Two-piece brass case with a maroon inner paper tube. Case length 65 mm; Bl 11 + 47; Pr MCI; St Ward & Son № 12 Worcester. (Inner) Kynoch Patent Grouse Ejector; Or British.

WARD & TAYLOR: Leominster, Herefordshire.
Business: Ironmongers.
Cartridges:
Example: Unnamed. Ga 12; Tc brown/black; Kynoch case; Or British. No other details recorded.

WARD THOMPSON BROTHERS: 87 Borough Rd, Middlesbrough, Yorks (Cleveland).
Cartridges: This name and address has been given to me but I do not have any information.

EDWIN WARING: High St, Leamington Spa, Warwickshire.
Business: Ironmonger and gunsmith.
Cartridges: Unnamed. Ga 12; Tc orange/black; Kn Specially loaded for Edwin Waring. Ironmonger and gunsmith. Telephone 555; Bl 8; Pr MCI; St ICI; Wd yellow/black. Smokeless · Diamond · 5; Or British.

JAS. B. WARRILOW: Chippenham, Wilts.
Business: Gunsmith.
Cartridges: Badminton; Ejector; Electric Long Shot Cartridge; The Good Sport.
Example: Badminton. Ga 12; Tc purple/silver; De A small crown; Kn His Grace, The Duke of Beaufort's Badminton Cartridge. Specially loaded. Smokeless. Jas B. Warrilow Chippenham (Copyright); Bl 11; Pr MCI; St J. B. Warrilow No 12 Chippenham. (Inner) Trade Mark Accurate; Wd light blue/black. J.B. Warrilow · Chippenham · 5½; Or Not known.

Mr J. B. Warrilow was in business in Chippenham between 1886 and 1913, or thereabouts.

WATKINS: Banbury, Oxon.
Cartridges:
Example: Unnamed. Ga 12; Tc dark blue/yellow; Kn Eley London; Bl 10; Pr SCI; St Watkins № 12 Banbury · Eley; Wd brown card/black. Watkins · Banbury · ; Or British.

JAS R. WATSON & CO: 35 Queen Victoria St, London. E.C.4.
Business: Gun and cartridge merchants.
Cartridges: The Britannia; The Challenge; Ejector; The Enterprise; The Lilliput; The Sureshot; The Warrior; The Wetteren.
Example one: The Challenge. Ga 12; Tc dark red/black; De Two opposing fighting cocks. This was a Registered Trade Mark; Kn Loaded with smokeless powder. Metal-lined case; Bl 8; Pr LCI; St Jas. R. Watson & Cº 12 12 London. (Inner) Made in Belgium; Wd green/black. (A crown) Cooppal Smokeless 5; Or Belgium.
Example two: The Warrior. Ga 12; Tc dark red/black; Kn British-loaded smokeless powder. Metal-lined case; Bl 8; Pr Not recorded; St Jas. R. Watson & Cº 12 12 London; Or Belgium.

This firm was established in 1889. Most of their brand cartridges were loaded into Belgian cases and used Cooppal powders. These powders were black powder Treble Strong and F. The smokeless sporting powders were No 1 and Emerald, these being granular. No 2 and Excelsior were both leaflet. Customers who ordered cartridges in quantities of 10,000 and upwards could have their own names printed on the tubes if they wished.

WATSON BROTHERS: Pall Mall, London.
Cartridges: Hi-speed Cartridge.
Example: Hi-speed Cartridge. Ga 12; Tc dark green/black; Bl 8; Pr MCI; St EN; Or British.

Case colour: Dark or forest green with black print.

WEBB: Hull, Yorks (Humberside).
Cartridges: An old cartridge head has been found with St Webb · Hull · Eley 12 London.

J. WEBBER & SONS: Exeter, Torquay & Newton Abbot, Devon.
Cartridges: Webber's ISCA.
Example: Webber's ISCA. Ga 12; Tc dark green/black; De A dark cock pheasant walking in grass with tail slightly raised, L; Kn Specially loaded with the highest-grade powder for J. Webber & Sons; Bl 16; Pr MCI; St ICI; Or British.

R. WEBLEY & SONS: London.
Cartridges: The remains of one old cartridge have been found with St R. Webley & Sons № 12 London.

G. R. WEBSTER: 30A Wide Bargate, Boston, Lincs.
Cartridges: The Favourite.
Example: The Favourite. Ga 12; Tc spring green/black; De A gliding partridge and banking over, L; Kn Specially loaded by G. R. Webster. Phone 3124; Bl 16; Pr MCI; St ICI; Wd yellow/black. Smokeless · Diamond · 4; Or British.

E. WEST: Retford, Notts.
Business: Gunmaker.
Cartridges:
Example: Unnamed. Ga 12; Tc dark red/black; De A cock pheasant standing in grass with his tail horizontal and the tip turned up, L; Kn All the tube wording is convexed and concaved being above and below the pheasant; Bl 8; Pr MCI; St E . West . Retford. Kynoch № 12; Or British.

TED WEST: Tetbury, Glos.
Business: Saddler.
Cartridges: Information given to me was that they once sold a cartridge with their name.

WEST & SON: 26 The Square, (also at 10 Bridge Gate St) Retford, Notts.
Business: Gunmakers.
Cartridges: The County; Ejector; Fourten; The Grand National; The Sherwood.
Example: The Grand National. Ga 12; Tc orange/black; Kn The name of the cartridge is in longhand. West & Son. Cartridge loaders. Retford. Telephone 66. Telegrams West, Gunmakers Retford; Bl 8; Pr MCI; St ICI; Wd pale orange/black. + West & Son + Retford. (Inner) Special Loading 4½; Or British.

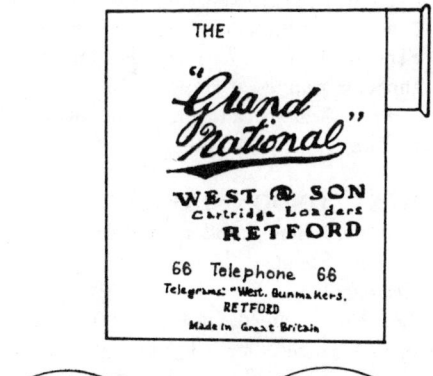

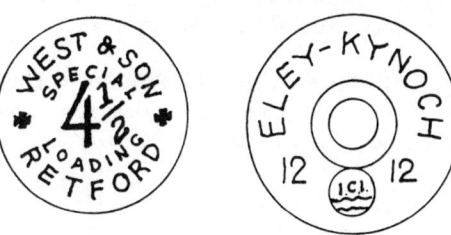

The Grand National. Case colour: Deep orange with black print. Wad colour as illustrated: Light orange with black print.

WEST & SON: Great Yarmouth, Norfolk.
Business: Gunsmiths.
Cartridges:
Example: Unnamed or name not known. Ga 12; Tc green. No other details recorded.

WESTLEY RICHARDS & CO: 12 Corporation St. (later at 24 Bennetts Hill) Birmingham. Also at 178 New Bond St. (Later at 23 Conduit St) London.
Business: Gunmakers.
Cartridges: The A.L.P. Cartridge; The Aquatite; The Carlton; The Explora Hollow Slug Cartridge; The Fauneta Ball Cartridge; Pegamoid; The Regent Metal Covered Cartridge; The Right & Left; The Special; The Wizard.
Example one: The Carlton. Ga 12; Tc light blue/black; De The triangle Trade Mark; Kn 24 Bennetts Hill. Shooting ground for gun fitting and tuition; Bl 16; Pr MCI; St KN; Wd white card/red. Westley Richards. Diamond Smokeless. (Shot size in a triangle) 6; Or British.
Example two: The Wizard. Ga 12; Tc orange/black; Kn 23 Conduit Street, London W.1. Telephone No Mayfair 5886. Telegraphic address, Hammerless, Piccy, London; Bl 8; Pr MCI; St ICI; Or British.

This firm moved from Corporation Street to Bennetts Hill in the November of 1910 and from New Bond Street to Conduit Street in the September of 1917. They also ran their own shooting grounds at Hendon, London W. Here they sold a special version of their Wizard cartridge.

WEST LONDON SHOOTING SCHOOL: Perivale, Ealing, London. W.
Business: Shooting School.
Cartridges:
Example: Unnamed. Ga 12; Tc purple/yellow; Bl 8; Pr LC-; St Nobel's Empire. (Inner) Made In Belgium · No · 12; Or Belgium.

CHARLES & HERBERT WESTON: The Colonnade, Brighton. Also at Hailsham. Later became C. & A. Weston.

C. & A. WESTON: The Colonnade, Brighton, Sussex.
Business: Gunmakers.
Cartridges: Brighton; Colonnade; Ejector; National Smokeless; Smokeless Cartridge; Special Smokeless.
Example one: C. & H. Weston's Special Smokeless Cartridge. Ga 16; Tc orange/black; Kn Brighton & Hailsham; Bl 6; Pr MCI; St Weston. Brighton & Hailsham. Kynoch № 16; Wd white card/purple. (Shot size only) 8; Or British.
Example two: C. & A. Weston's Brighton Cartridge. Ga 12; Tc crimson/black; De Crested coat of arms; Kn Brighton; Bl 16; Pr MCI; St ICI; Wd yellow/black. Smokeless · Diamond · 5; Or British.

Charles and Herbert Weston were trading in the gunmaking business on the Colonnade at Brighton since the start of breechloading guns. Their cartridges were marked C. & H. Weston. In later years the firm traded as C. & A. Weston's. The business terminated during the 1970s.

J. WHEATER: 7 Queen St, Hull, Yorks (Humberside).
Business: Gunmaker.
Cartridges: The Humber.
Example: The Humber. Ga 12; Tc orange/black; De A duck taking off over reeds, R; Bl 8; Pr MCI; St ICI; Or British.

J. Wheater took over the business from R. Robinson who also marketed The Humber cartridge.

J. E. WHITEHOUSE: Oakham, Rutland (Leics).
Cartridges: The Quorn; The Rutland.
Example: The Quorn. Ga 12; Tc purple/silver; Bl 8; St Whitehouse № 12 Oakham: No other details recorded.

F. S. WHITEMAN: Wallingford, Berks (Oxon).
Business: Ironmonger and gunsmith.
Cartridges: The Fordian.
Example: The Fordian. Ga 12; Tc maroon/gold; De A hen pheasant walking with her tail high, L; Kn Special smokeless. British hand loaded. Ironmonger. Gunsmith. Phone, 76; Bl 7; Pr –CI; St Special 12 12 Smokeless. (Inner) Foreign Made Case; Wd pale yellow/black. Shot size only being a large figure 5 with a ball tail; Or Foreign.

WHITEMAN BROTHERS: Silver St & The Tything, Worcester, Worcs.
Cartridges: The Defiance; The Rapid.
Example: The Rapid. Ga 12; Tc red/black; De An overhead gliding partridge, Fr; Kn Specially loaded by Whiteman Bros L$^{td.}$ Silver Street, Worcester. Entirely British made; Bl 8; Pr MCI; St ICI; Wd white card/black. (Shot size only) BB; Or British.

WICHELL: Tetbury, Glos.
Business: Ironmonger.
Cartridges: Information given was that they once sold cartridges.

THOMAS WILD: Birmingham (W. Midlands).
Cartridges: Remington New Club.
Example: Remington New Club. Ga 12; Tc yellow/black; Bl 8; Pr MCI; St REM-UMC № 12 New Club; Or U.S.A.
  I presume that Thomas Wild must have had his name on the cartridge in some place.

JOHN WILKES: 79 Beak St, London. W.1.
Cartridges: Tom-Tom.
Example: Tom-Tom. Ga 12; Tc red/black; Bl 8; Pr MCI; St ICI; Wd yellow/black. Special · Smokeless · 6; Or British.

WILKINSON: Pall Mall, London S.W.
Business: Gun and sword makers.
Cartridges: Regal; Special.
Example: Regal. Ga 12; St Wilkinson Pall Mall. No other details recorded.

WILKINSONS: Sports Depot, Penrith, Cumberland (Cumbria).
Business: Sports depot.
Cartridges: The Eden.
Example: The Eden. Ga 12; Tc Off-white/blue; Bl 8; Pr MCI; St ICI; Or British.

WILKINSON'S: Durham, Co. Durham.
Business: Ironmongers.
Cartridges:
Example: Unnamed. Ga 12; Tc red/black; De A coat of arms consisting of a dark shield with two light bars forming a cross on it. A half wreath of leaves decorate each side of the shield and join at the bottom; Kn Specially loaded by Wilkinson's (Late J. R. Malcolm). Ironmonger. Durham; Bl 8; Pr MCI; St KB; Wd light yellow/black. Wilkinsons · Durham · 7; Or British.

J. W. WILLCOCKS: Stamford, Lincs.
Cartridges: Eley's Gas-tight Cartridge Case for E.C. Gunpowder.
Example: The above. Ga 12; Tc Brownish tan/black; Kn The above printed diagonally; Bl 10; Pr SCI; St J. W. Willcocks . Stamford. · № 12 Eley; Wd white card/red. Willcocks · Stamford · 5; Or British.

C. D. WILLIAMS: Belfast, Northern Ireland.
Cartridges: Their name has been seen on an old cartridge head.

HARRY WILLIAMS: 49 Pyle St, Newport, Isle of Wight.
Business: Ironmonger.
Cartridges: The Express.
Example: Unnamed. Ga 12; Tc dark red/black; De A cock pheasant standing with his tail out horizontal, L; Bl 8; Pr MCI; St KB; Wd white card/red. Harry Williams. Newport. I.o.W. 4; Or British.

J. S. WILLIAMS: Pontypridd, Glam.
Business: Ironmonger and explosives merchant.
Cartridges:
Example: Unnamed. Ga 12; Tc mauve/black; Kn J. S. Williams. Ironmonger and explosives merchant. Pontypridd; Bl 7; Pr MCI; St J. S. Williams № 12 Pontypridd; Wd white card/black. Special · Smokeless. 7; Or Not known.

WILLIAMS & POWELL: Liverpool, Lancs (Merseyside).
Business: Gun, pistol and rifle makers.
Cartridges: The Castle; Kynoch Patent Grouse Ejector.
Example one: Unnamed. Ga 16; Tc Waxed turquoise/black; Kn Waterproofed Gastight cartridge manufactured by F. Joyce & Co, London. Bl 10 (Bronze); Pr Not recorded; St Williams & Powell. L'pool. № 16; Or British.
Example two: The Castle. Ga 12; Tc red/black; Kn Smokeless. Williams & Powell. Gunmakers. Liverpool; Bl 8; Pr MCI; St KB; Or British.

C. WILLIAMSON: Stockton-on-Tees, Co. Durham (Cleveland).
Cartridges:
Example: Unnamed. Ga 12; Tc dark blue/nil; Bl 10; Extra thick rim; Pr Not recorded; St C. Williamson . Stockton-on-Tees. № 12 Eley; Or British.

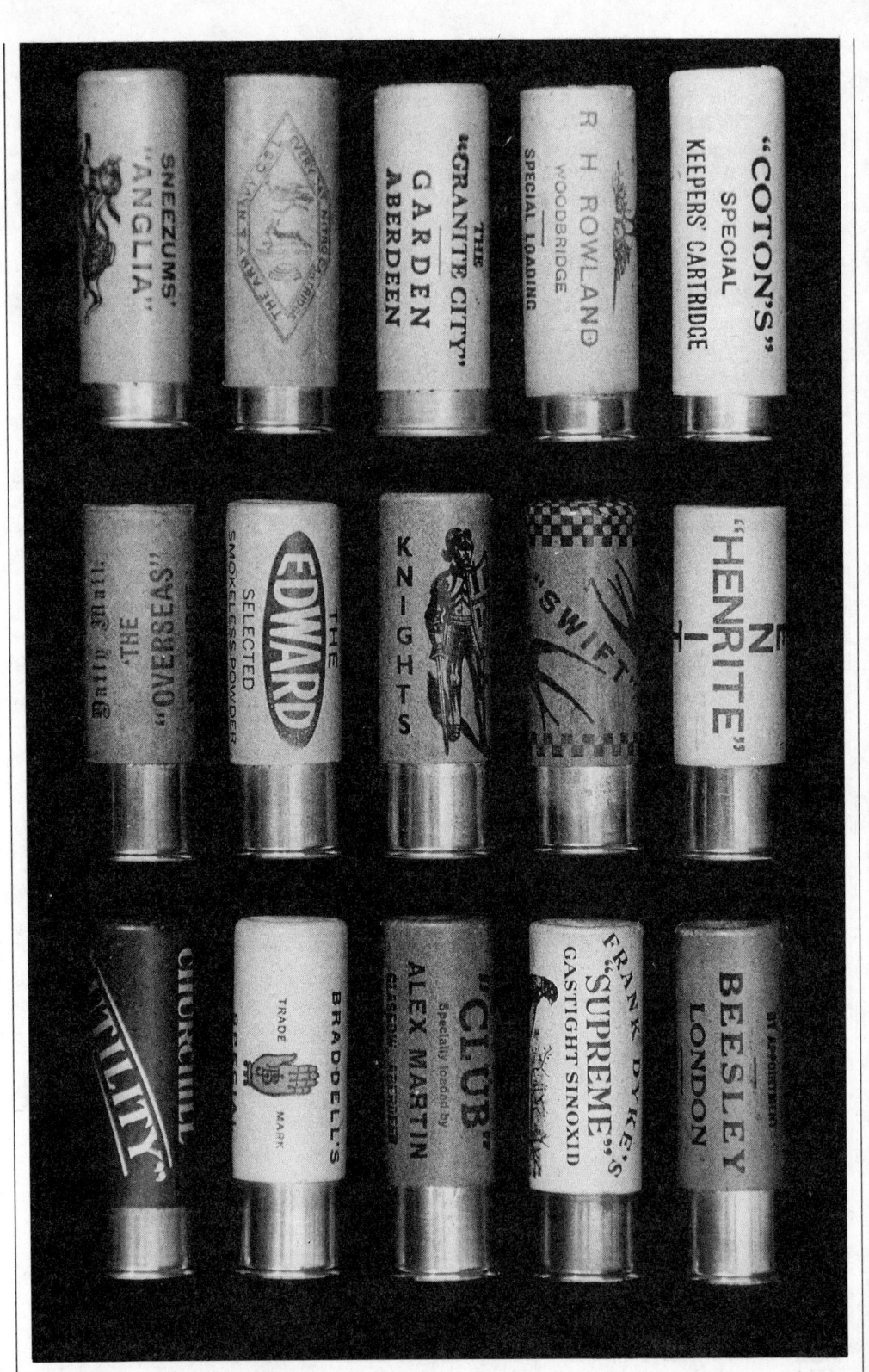

A selection of famous cartridge names

An impressive Eley cartridge display

D. WILLIAMSON: 5 Waterloo Bridge Rd, London.
Business: Gunmaker.
Cartridges: Ejector.
Example: Ejector. Ga 12; Two-piece brass case with a maroon inner paper tube; Bl 8 + 48; Pr Not recorded; St E 12 12; Wd white card/blue. Williamson · London · 6; Or British.

WILSON: Norwich, Norfolk.
Business: Gunmaker.
Cartridges: I regret that I have no cartridge information.

G. H. WILSON: 9 Market Place, Horncastle, Lincs.
Business: Gunmaker.
Cartridges: The Champion.
Example: The Champion. Ga 12; Tc light brown/black; Bl 11. No other details recorded.
This firm was taken over by Mr A. Hill at the same address in 1902.

J. WILSON: Later as J. Wilson & Son; York, Yorks.
Cartridges: Ejector.
Example: Ejector. Ga 16; Two-piece brass case with a brown inner paper tube; Bl 10 + 47; Pr MCI; St Eley's Ejector · Nº 16 London; Wd white card/pink. J. Wilson · York · 7; Or British.
The following top wad has also been seen: Wd yellow/black. J. Wilson & Son · York · 4.

WOOD: Salisbury, Wilts.
Cartridges:
Example: Unnamed. Pinfire. Ga 12; Tc brown quality/nil; Bl 8; Pr Pinfire; St KB; Wd Their name was on the wad only; Or British.

ARTHUR WOOD (NEWPORT. I.W.): 114 Pyle St, Newport, Isle of Wight.
Cartridges:
Example: Unnamed. Ga 12; Tc maroon/gold; De Cock pheasant walking on pine needles, L; Kn Special smokeless; Bl 8; Pr MCI; St SS; Wd light yellow/black. Special · Smokeless · 6.; Or Foreign.

GEORGE WOOD & CO: Sheffield, Yorks.
Business: Gunsmiths.
Cartridges: I do not have any cartridge information, but this business was absorbed into that of H. B. Suggs of Nottingham.

J. L. WOOD: Stamford, Lincs.
Cartridges:
Example: Unnamed. Ga 12; Tc dark blue/nil; Bl 11; Pr SCI; St J. L. Wood Nº 12 Stamford; Or Not known.

WOOD & HORSPOOL: Newport, Isle of Wight.
Cartridges:
Example: Unnamed. Ga 12; Tc green/nil; Bl Short brass; Pr SCI; St 12; Or Not known.
A Ga 12 pinfire has also been seen by them. This business was terminated in the 1880s and later carried on by Arthur Wood.

WOODDISSE & DESBOROUGH: Ashbourne, Derbyshire.
Cartridges: The Premier.
Example: The Premier. Ga 12; Tc light brown/black; De The EBL shield Registered Trade Mark; Bl 10; Pr MCI; St ELG; Or British.

WOODROWS: Salisbury, Wilts.
Business: Ironmongers.
Cartridges: Information given to me was that they once loaded cartridges.

GEORGE L. WOODS: Also, **G. L. WOODS & SONS**: Ovington, Norfolk.
Business: Cartridge loaders and gunsmiths.
Cartridges: Castle Forbes; Norfolk Universal; Woods Special; Woods Supreme.
Example: Castle Forbes. Ga 12; Tc orange/black; Bl 8; Pr MCI; St MGB; Or British.
Geo L. Woods have loaded many different cartridges and some with brand names not listed above as I have only seen them with cases that have been closed with the star crimp. They have also loaded to private order such as Woburn R.C.S. Bedfordshire.

JAMES WOODWARD & SONS: 64 St James's St, Pall Mall, London. S.W.
Business: Gunmakers.
Cartridges: The Automatic; Kynoch Patent Grouse No 2090 Ejector.
Example: Unnamed. Ga 12; Tc gastight brick red/black; Kn Water resisting. Steel lined; Bl 16; Pr MCI; St ICI; Wd white card/black. Woodwards -·- 7; Or British.

GEORGE WREN: High St, Hungerford, Berks. Also at Ramsbury, Wilts.
Business: Ironmonger.
Cartridges:
Example: Unnamed. Ga 12; Tc light brown/black; De A shield coat of arms with a scroll below with the name Hungerford on it; Kn G. Wren. Hungerford and Ramsbury; Bl 16; Pr MCI; St KB; Wd red/black. G. Wren · Hungerford · 5; Or British.

The example is the only cartridge of his that I have ever seen although I did hear of a pinfire that had his name on the top wad. The business changed hands in 1925.

JAMES WRIGHT: Okehampton, Devon.
Business: Ironmonger.
Cartridges: Special Load.

A. B. WYLIE: 64, 66 Market Place, Warwick, Warwickshire.
Business: Ironmonger.
Cartridges: At one time they did load their own cartridges and many years ago I was taken upstairs and shown the old loading machine. Alas, there were no old cartridges.

Some Unidentified Cartridges

MANY of the firms mentioned in the list, although they had their names on the cartridges, did not always give them brand names. In the majority of cases their names were printed on the paper tubes, but this was not always the case. In some instances it was only on the headstampings or the overshot wads. The firm's name has not always been printed on the cartridge; sometimes the name has only been carried on the box in which they were sold. I have seen unopened boxes of cartridges with a firm's name and a brand name printed on the box lid and when the box has been opened the cartridges inside had no printing on their tubes. This was quite common practice during the two world wars.

Below is a selection of cartridges to which I have been unable to put a firm's name. Although I have assumed they are from the British Isles, this is not certain.

20 Gauge 20. Smokeless Gastight
De Luxe. Water Resisting. Loaded in Ireland
English Pioneer Cartridge
The Express Special Smokeless
Gastight Metal Lined Cartridge
M. & C. CA. (on headstamp only)
Metalode Metallic Smokeless
The Rabbit Special

Smokeless Cartridge
Smokeless Metal Lined Cartridge
Special British
The Supreme
The Trojan
The Union Jack Brand Cartridge
The Windsor Cartridge

An Australian Collection

As the Australian market was supplied for many years by Britain, I think it is only proper to include information that I have gleaned whilst I have been in Australia. This has been made possible through the kindness of Ken Mitchell and Geoff Shawcross, who have given me access to their own records.

Most of the Australian cartridge activity seems to have taken place in the south-east corner of the country. The few firms and their cartridges that I am listing here are only a portion of the many firms in Australia that once sold cartridges bearing their own names.

An additional abbreviation used in this Australian List.

MIA

ALCOCK & PIERCE: 318 Little Collins St, Melbourne, Victoria.
Business: Field sport stores.
Cartridges: The A.P.; Apoid; Blue Seal; Ebnoid; Kilos.
Example: Blue Seal. Ga 12; Tc grey/dark blue; De A seal with the print forming a background for the name Blue Seal in grey; Kn 318 Lt Collin St Melbourne; Bl 8; Pr MCI; St MIA; Wd orange/black. Smokeless -·- BB; Or Australian.

ALDERSON'S: Sydney, New South Wales.
Cartridges: Algame Express; Algame Express High Velocity.
Example: Algame Express. Ga 12; Tc cream or grey/black; Kn The name on the cartridge was printed diagonally in longhand; Bl 8; Pr MCI; St MIA; Wd white card/black. Chilled Shot × 3; Or Australian.

AMMUNITION (NOBEL) PTY, Ltd.
See I.C.I.A.N.Z. in this list.

BRISCOE & CO: Melbourne, Victoria & Sydney, New South Wales.
Business: Merchants
Cartridges: Briskill; Essex.
This firm was at one time the agents for marketing Nobel Explosives cartridges in Australia. All of their own brand cartridges were loaded by I.C.I.

S. A. CASE: 377 Little Collins St, Melbourne Victoria.
Business: Cartridge loader.
Cartridges: Case-loaded cartridges with Amberite Powder and his boxes were labelled, 'Case's Cartridge'. Mr Case blew himself to pieces in his own factory when opening a keg of powder while smoking his pipe! Circa 1915.

CASTES BROTHERS: Kyneton, Victoria.
Cartridges:
Example: Unnamed. Ga 12; Tc orange/nil, Bl 10; Pr MCI; St MIA; Wd white card/black. Castes Bros-Kyneton · 12; Or Australian.

CHANDLER'S: Melbourne, Victoria.
Cartridges: Kangaroo Black Powder Load; Kangaroo Smokeless.
Example: Kangaroo Smokeless. Ga 12; Tc buff/black; Kn Smokeless powder cartridge; Bl 8; Pr MCI; St KB; Wd white card/black. 6 Smokeless 6; Or British.

FRED CLIFT: Little Bourke St, Melbourne, Victoria.
Business: Gunsmith.
Cartridges: The Field.
Fred was one of the first cartridge loaders in Victoria. His Field cartridges were sold in an attractive box. They were loaded with clear celluloid overshot wads. Headstamps were Eley Gastight of British origin.

COWLES & DUNN: Sydney, New South Wales.
Cartridges: Eley E.B.; Eley Grand Prix; Kynoch Unlined Nitro Case; Nitrone.
Example: Eley Grand Prix. Ga 12; Tc orange/black; De The EBL shield Registered Trade Mark; Kn Eley Brothers 'Grand Prix' printing; Bl 8; Pr MCI; St EN; Wd white card/red. Cowles & Dunn · Sydney · 6; Or British.
This firm retailed the normal Eley and Kynoch brand cases loaded with their own overshot wads.

W. E. EKINS & CO: Adelaide, South Australia.
Business: Cartridge loaders and merchants.
Cartridges: Ballistite Cartridge; Eclipse; Rabbird; Special Duck Load.
Example: Eclipse. Ga 12; Tc orange/black; Kn Loaded by W. E. Ekins, Adelaide. Case made in England; Bl 8; Pr MCI; St EN; Wd white card/black. Loaded by W. E. Ekins · 2; Or British.
W. E. Ekins & Co, were first established in 1860 in King William Street. They later moved to 29 Hindley Street and then on to 92 Currie Street. All these moves were in Adelaide. Some of the first cartridges they loaded were pinfires. Cartridge cases, or shells as the Australians call them, were imported from England, France, Germany, Belgium and Austria. Some of the British cases they loaded were the Eley Grand Prix and the Gastight. These were prior to 1936 before Australia manufactured its own shells. Their own brand cartridge called the Rabbird was derived from a combination of the words rabbit and bird. During their years of business Ekins bought out the firm of James Green & Son. Packeted label codes for their shot shells were Orange, Black powder; Blue, Green and Red, Smokeless powders. Ekins finally closed their doors in 1969.

ELEY BROTHERS:
See I.C.I.N.Z. in this list.

W. J. FELL: Warracknabeal, Victoria.
Business: Tinsmith and gun trader.
Cartridges: A cartridge head was found with St W. J. Fell. № 12 Warracknabeal.

GORDON CARTRIDGE CO: Gordon, Sydney, New South Wales.
Business: Cartridge loaders.
Cartridges: Bull's Eye; Gordon Stars.

HARTLEY'S SPORTS STORE: Flinders St, Melbourne, Victoria.
Business: Sports Store.
Cartridges: Diamond; Hartsport.
Example: Hartsport. Ga 12; Tc crimson/black; Kn Long range. (All tube printing is diagonal); Bl 8; Pr MCI; St MIA; Wd white card/black. Chilled Shot 4; Or Australian.

HARVEY SHAW SUCCESSORS: 560-6 Lonsdale St, Melbourne, Victoria.
Business: Cartridge loaders and merchants.
Cartridges: Rhenwest.
Example: Rhenwest. Ga 12; Tc trainer yellow or kingfisher blue/black; Kn loaded with Walsrode; Bl 8; Pr MCI; St Loaded in 12 12 Australia. (Inner) Shell made in Germany; Wd vermilion/black. Field · Load · 3; Or German.
Harvey Shaw Successors were formed in the early 1930s to load Rhenwest cases made by Rhenn West P/L in Germany and using German Walsrode Powder. They were a subsidiary of D. W. Chandler, who distributed their cartridges. Principals were R. Aylett and J. M. Allen. Heavy loads were loaded into high brass shells with yellow or red paper tubes. Field loads were loaded into low brass shells with yellow or blue paper tubes. Quail loads were loaded into low brass shells with yellow or red paper tubes that were marked 'Specially waterproofed for duck shooters'. Cartridges were loaded in Ga's 12, 16, 20 and ·410. Sinoxid primers were used. Shells were also loaded for the wool broker Henry B. Smith in both high and low brass with pale green paper tubes. Also, shells were loaded for Pay's Stores and the New Zealand Loan & Mercantile Co Ltd. All Australian Rhenwest cartridges are pre-Second World War, as the war put paid to any future activity.

HAWKE'S BROTHERS: Geelong, Victoria.
Cartridges: Nobel's Corio; Barwon; Moorabool.
Example: Nobel's Corio. Ga 12; Tc not recorded; Kn Nobel's Corio Smokeless Cartridge. Made in Great Britain; Bl 8; Pr MCI; St NGN; Wd white card/black. Chilled shot × 12; Or Scottish. The above example was loaded for Hawke's brothers by Nobel Explosives Ltd in Scotland. The name Corio being taken from Corio Bay, Geelong.

HOFFNUNGS: Sydney, New South Wales.
Cartridges: Straightline; Super Straightline; Texas Jack.
Example: Straightline. Ga 12; Tc dark green/black; Kn Schultze Powder. Chilled shot. Made in Great Britain; Bl 8; Pr MCI; St ICI; Wd orange/black. Schultze Powder * 4; Or British or Australian.

ANTHONY HORDERNS: Brickfield Hill, Sydney, New South Wales.
Business: Clothing and sports store.
Cartridges: Tree Brand.
Example: Tree Brand. Ga 12; Tc not recorded; De A stubby tree with the word Tree on the left and the word Brand on the right. Below this a scroll has the wording, 'While I Live I'll Grow'; Bl 8; Pr MCI; St EL; Wd colour not recorded. Eley · Smokeless · 3; Or British.

I.C.I.A.N.Z. (IMPERIAL CHEMICAL INDUSTRIES OF AUSTRALIA & NEW ZEALAND): Deer Park, Melbourne, Victoria. I am here also including early Eley, Kynoch and Nobel loaded cartridges.
Business: Cartridge manufacturers and merchants.
Cartridges: 20 Gauge; 16 Gauge; Amberite; Blank; Blue Star; Bonax; Clyde; Coralite; Corio; Duxbac; Eley Unlined Nitro Case; Empire G–R; Empress; Fourlong ·410; Fourten ·410; Grand Prix Cartridge; Grand Prix Case; Heviload; Icil Special; Kynoch's Unlined Nitro Case; Leaderite; Maximum; Nile; Nitrone; Reliance; Rex; Rocket; Scare Bird; Special Quail; Special Trapshooting; Sporting Ballistite; Swift; Universal; Victor.

What makes the listing of the above cartridges very difficult is that some have the names Eley, Kynoch or Nobel on them while others have no names at all. It is also nearly impossible to identify some early Australian loads from some British loads. Also the headstamp Eley-Kynoch ICI ran for two separate periods and was used by England and Australia. Some early cartridge brands that were loaded in England were made solely for the Australian market. I believe this to be true of the Eley Brothers' Coralite and the Leaderite. Also, the Nobel's Victor. The Empress, Heviload, Rex, Special Quail and the Swift are all of Eley origin. The Corio (See Hawke's Brothers) and the Reliance are both Nobel brands. I have not listed the Eley's Comet and the ICIL Magnum as I have only seen them with a star crimp closure.

Example One: Empress. Ga 12; Tc grey/black; De The latter type large EBL shield Registered Trade Mark; Kn Made and Loaded in Great Britain; Bl 8; Pr MCI; St EN; Wd Not recorded; Or British.

Example Two: Eley Unlined Nitro Case. Ga 12; Tc middle blue/black; De A continuous diamond marking around the top of the paper tube; Kn loaded with Curtis's & Harvey's Smokeless Diamond Powder. Manufactured and loaded in Australia by Imperial Chemical Industries of A. & N.Z. Ltd; Bl 8; Pr MCI; St MIA; Wd orange/black. Smokeless -·- 10; Or Australian.

Example Three: Nobel Nile. Ga 12; Tc orange/black; Kn Sporting Ballistite Powder. Loaded in Australia; Bl 8; Pr MCI; St EN; Wd white card/black. Chilled Shot × 4; Or British.

Example Four: Eley 20 Gauge Case. Ga 20; Tc buff/black; De The later type large EBL shield Registered Trade Mark; Kn, Manufactured and Loaded in Australia by Imperial Chemical Industries of A. & N.Z. Ltd; Bl 7; Pr MCI; St ICI; Wd white card/black. Smokeless * 4; Or Australian.

Prior to World War One, Australia was served with loaded cartridges from Great Britain. Eley Brothers had an agent in Sydney who marketed their products over the entire country. G.. Kynoch & Co established a branch office in 1905 at Melbourne and in 1912 they started work on their own factory at West Footscray, Melbourne. Hand-loading equipment was installed and imported components were assembled. These works then also serviced gas-engines that Kynoch had supplied to Australia. The Nobel Explosives Company also had a representative in Australia and their cartridges were sold through their agent Briscoe & Co Ltd. Later they changed their agency to Dalgety & Co Ltd. Both of these firms operated out of Melbourne and Sydney.

About the same time that Kynoch's had

started work on their factory, another factory was being built at the Australian Explosives Chemical Company's Deer Park Works, where Nobel blasting explosives used for mining were already under manufacture. This new factory was sited not far from the Kynoch factory at West Footscray. Automatic loading machines were installed and Nobel and Eley brand shotgun cartridges were loaded.

After the war the Australian shotgun cartridge loading got under way again in the old Kynoch works at West Footscray as the gas-engine side of the business was then on the decline. In 1924 this operating company became officially known as Ammunition (Nobel) Pty, Ltd. New loading machines had been installed and the factory loaded the cases that were shipped out from the parent firm at Witton, Birmingham, England. These cases then had the same headstampings and often the same tube printing as that used on the home market.

During 1928 the I.C.I.A.N.Z. (Imperial Chemical Industries of Australia and New Zealand) was formed and in 1935 work was started on a new factory at Deer Park. A year or so later complete cartridge manufacture got under way. Then completed shells with percussion caps were made. Apart from the normal centre-fire, rimfire was also made. From then on I.C.I.A.N.Z. was to be seen on headstamps. The near-by West Footscray factory was put into use for storage purposes only.

With the coming of the Second World War this factory then became fully occupied with producing munitions. This included paper-tubed signal cartridges. After the war this factory went back to producing sporting ammunition and engine starter cartridges. During 1960, a 180 ft (56 m) high shot tower was built. From then on it supplied all the factory's requirements with lead shot. As I.C.I.A.N.Z. it carried on manufacturing ammunition until April Fool's day in 1970. On this date the controlling interests were taken over by Imperial Metal Industries (Australia) Ltd. The shotshells that were in production at the time were in Ga's 12, 16, 20 and ·410. The ·22 rimfires were also being manufactured.

IMPERIAL METAL INDUSTRIES (AUSTRALIA):
See I.C.I.A.N.Z. in this list.

G. KYNOCH & CO:
See I.C.I.A.N.Z. in this list.

J. T. LAKE: Adelaide, South Australia.
Business: Gunsmith.
Cartridges: The Lake Cartridge.
  The Lake was loaded for him by G. Kynoch & Co.

LOCKSLEY CARTRIDGE CO: Coogee, Sydney, New South Wales.
Business: Cartridge manufacturers.
Cartridges: Dead Bird; Killer; Royal Reserve; Settler; Special Pigeon.
  The Locksley Cartridge Co were in business between 1920 and 1935.

JAMES McEWAN & CO: Elizabeth St, Melbourne, Victoria.
Business: Agricultural and hardware merchants.
Cartridges: Padlock.
Example: Padlock. Ga 12; Tc mauve/yellow; De A padlock with J. McE. & Co. within. Kn Special Smokeless Cartridge; Bl 8; Pr MCI; St NG (Square-type lettering); Wd Not recorded; Or Scottish.

DONALD MACKINTOSH: Melbourne, Victoria.
Business: Gunmaker or gunsmith.
Cartridges: 3-inch Mackintosh Special Match.
  His overshot wads have been seen loaded into Eley, Kynoch and Remington–UMC shells.

MELBOURNE SPORTS DEPOT:
Melbourne, Victoria.
Business: Sports stores.
Cartridges: Meteor.
Example: Meteor. Ga 12; Tc orange/black; Kn Hand loaded in England for Melbourne Sports Depot. Cartridge Case Made By Eley-Kynoch (Around the tube just above the brass head); Bl 8; Pr MCI; St ICI; Wd Not recorded; Or British.

J. MUES: Post Office Place, Melbourne, Victoria.
Business: Cartridge loader and merchant.
Cartridges: Blackshell Long Range; Blackshell Trap Load; Redshell Falcon; Redshell X-Ray. Some US cases were Climax Smokeless; Defiance Smokeless; Ajax Heavies Long Range.
Example: Blackshell Trap Load. Ga 12; Tc black/silver; Kn Sporting Cartridge. Water resisting. Metal lined; Bl 16; Pr MCI; St J. Mues 12 12 Melbourne; Wd Not recorded; Or Not known.

Jack Mues started his own business in Melbourne in 1905 and produced his own brands of shotshells. He often used Remington-UMC cases, which included Nitro-Club and Arrow. He later imported Belgian F-N cases and also had his name on the headstamps. He was joined in the business by his son Charles and together they imported black US Climax cases. These had the headstamp Climax Shell № 12. Made in U.S.A. Loaded in Australia.

Later I.C.I.A.N.Z. loaded cartridges for them, but they had to use special cases that had black paper tubes. All their early red-tubed cartridges had black printing and their black-tubed cartridges had silver printing. Jack Mues died during the Second World War and Charles carried on the business until after the war. The business was then sold to Horrie James, who was a town shotgun shooter and shot under the name of Duxbac. This name had first been given to a popular pre-war ICI cartridge.

NOBEL EXPLOSIVES CO:
See I.C.I.A.N.Z. in this list.

PAY'S STORES: Kerang, Victoria.
Business: Large stores.
Cartridges: Rhenwest.
Example: Rhenwest. Ga 12; Tc crimson/black; Kn Loaded with Walsrode. It pays to buy at Pay's (Additional tube printing); Bl 8; Pr MCI; St Loaded in 12 12 Australia. (Inner), Shell Made In Germany; Wd orange/black. Field · Load · 4; Or German.
Pay's Stores closed down at about the time of the coming of the Second World War. For further information see Harvey Shaw Successors in this list.

J. W. ROSIER: Bourke St, Melbourne, Victoria.
Cartridges: I could not find any information on his cartridges. His name has been seen stamped on the bottom of an old cartridge head.

R. F. SCOTT & CO: Ballarat, Victoria.
Cartridges: I have been shown the following overshot wad. Wd white card/red. R. F. Scott & Co · Ballarat.

MICK SIMMONDS: Hay Market, Sydney, New South Wales.
Business: Sports stores.
Cartridges: The Lightning; Special Smokeless; Super Lightning; Warren.
Example: The Lightning. Ga 20; Tc middle blue/black; De A picture within a square frame of three jumping kangaroos, L. A fallen 'roo is in the bottom right-hand corner and a hunter on horseback is in the top right-hand corner; Kn Smokeless. High Velocity. Standard Load. (All the wording is within a frame); Bl 7; Pr LBI; St Smokeless 20 20; Wd white card/black (Shot size only) 4; Or French.

Some unidentified Australian cartridges

The Ak-U-Rate. Ga 12; Tc orange/black; Kn Smokeless Cartridge. Sure & Deadly; Bl 8; Pr MCI; St ICI; Or Australian or British.

Boomerang. Ga 16; Tc eucalyptus green/nil; Bl 5; Pr MCI; St The Boomerang 16; Wd Not recorded; Or Not Known.

Top Brand. Ga 12; Tc kingfisher blue/black; De A child's conical wooden-type top with the words within and at its top, 'Top Brand'. This is a Registered Trade Mark; Kn Smokeless Cartridge. Loaded in Australia; Bl 8; Pr MCI; St EN; Wd Not recorded; Or British.

An unknown firm in Adelaide, South Australia, has been reported to have loaded the following brand-named cartridges: Enterprise; Mullerite Setter; Phar Lap.

Australian Cartridges Post Second World War

APART from I.C.I. and I.M.I.'s activity, which has diminished following the Second World War, the Super Cartridge Company Limited of Melbourne now manufactures its own shells. These were made in one-piece all alloy and later in one-piece all polythene plastic, although paper has been used. The plastic Ga 12 shells have been made in many colours including black, grey and white. In their early days they imported some red paper Italian Fiocchi cases and loaded them into a cartridge which they named the Super Express.

The firm of Australian Cartridge Enterprises Limited started up in business in Melbourne. They also marketed a Ga 12 cartridge loaded into a red paper Fiocchi case. This they called the Ace, the cartridge taking its name from the first letters in the firm's name. This cartridge was closed by a star crimp. The firm only ran for a few years and on the collapse of the business, one man who held some controlling interests then started up in business on his own. His name is George Biggs and he now produces his own cartridges loaded into plastic shells.

The Australian cartridge market was further swollen when the American firm of Winchester–Western produced their own plant at Hays Road, Geelong, Victoria, and captured the Australian market under the name of Winchester Division, Olin, Australia Limited. With plastic-tubed shells their cartridges are loaded with the headstamping Winchester 12 Ga Australia. They now manufacture a variety of brands. Some of these are Light Game, Quail Load, Rabbit Load etc. Also loaded are cartridges for special clay shoots.

The Colonial Ammunition Company Limited (C.A.C.), associated with New Zealand, also had large sales in Australia. This firm dates so far back that some of their first cartridges were to be found with Eley Brothers' E.B. mark on the stampings along with the wording of the C.A.C. This firm is now absorbed into the I.M.I. complex.

Names of Some Firms to be Found Within the Cartridge

THE BREAKABLE WAD CO: Newcastle-upon-Tyne, Nthmb (Tyne & Wear).
This firm manufactured their patent breakable top wads. These were advertised in 12-, 16- and 20-bores at 5/- per 1,000 to the rear of Eley Brothers' price list for 1902. Possibly for other years as well.

Wads above were white card/black

CL
Patent shot concentrator. The one in my collection consists of a 13 mm length of brown card tube in 16-bore. Tc middle green/black. De A Trade Mark, this being a triangle with the letters CL in its centre; Kn Patent Shot Concentrator. I have no other details.

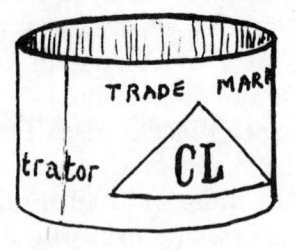

CL Shot Concentrator

AUTOMATIC SHRAPNELL CO: Edinburgh, Midlothian.
This firm produced the Johns Sporting Shrapnell Shell in several different variations. These were patented from the invention by James Watson Johns of White House, Cramond Bridge near Edinburgh. These shells consisted of a brass nutshell-type canister that contained the shot load. The brass canister was made in two halves and was held together by a wire spindle that protruded upwards from a card wad below. These brass pressings have holes in their tops and bottoms for the spindle to poke through. The shape of the containers or canisters depended upon the weight of shot load used. These Shrapnell Shells have been seen loaded into some cartridges before the turn of the last century and it was claimed that they held the shot in a concentrated cluster for up to 90 yards (approx 80 m), giving bird kills as far as 130 yards (approx 120 m).

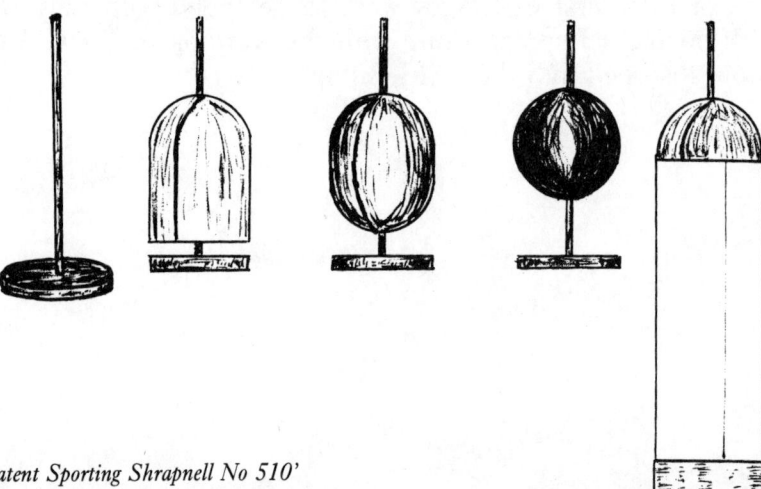

'Johns Patent Sporting Shrapnell No 510' was stamped on the canister sides

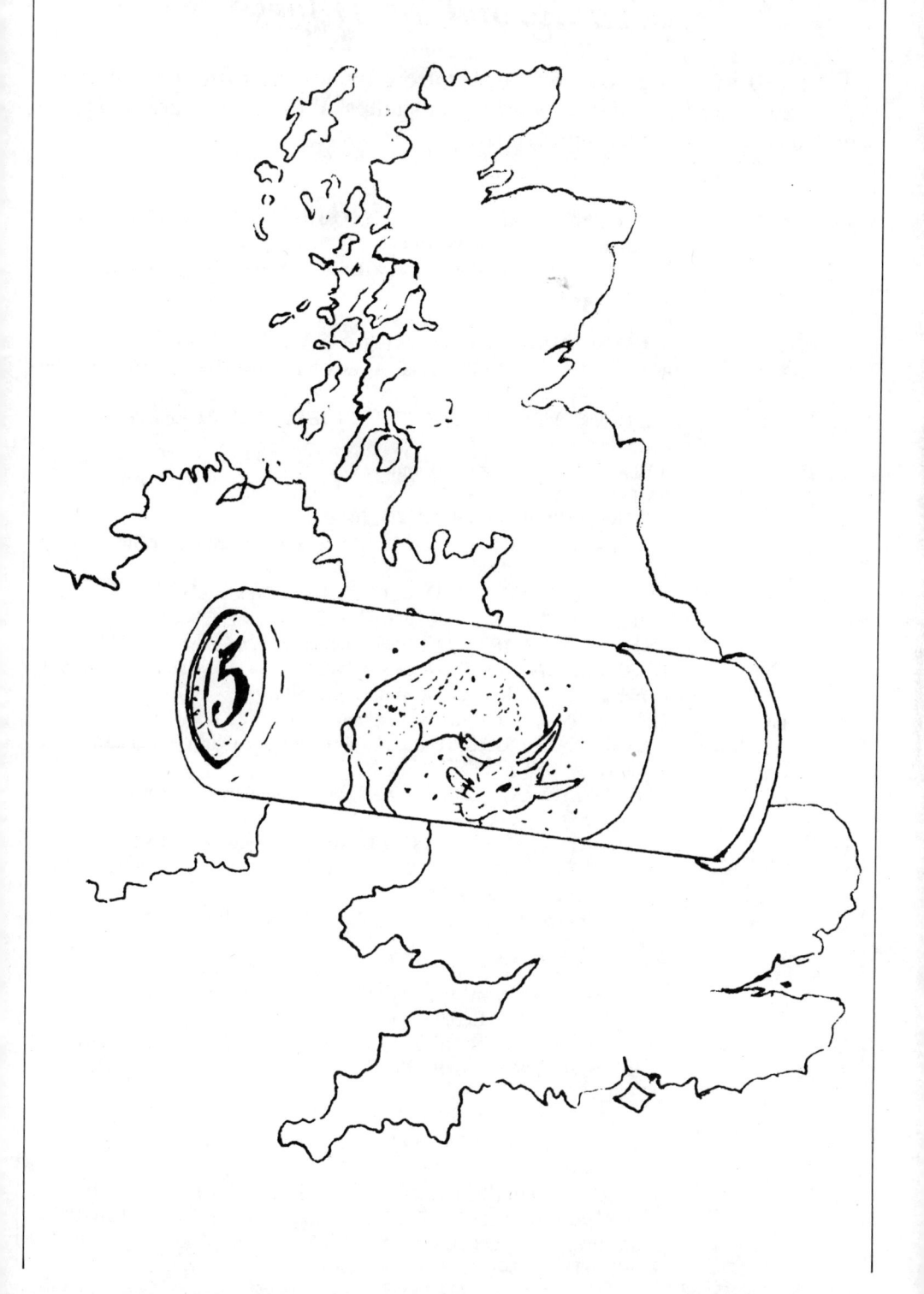
5

Headstamp Letter Combinations on Shotgun Cartridges

LISTED here in alphabetical order are a few of the many letter combinations to be found on world headstamps. Where dates are shown in brackets, they are only approximate.

A.A. CO.	AMERICAN AMMUNITION CO: Oak Park & Chicago, Ill. Also at, Muscatine, Iowa. U.S.A. (1910 onwards. Not known when ceased).
A.B.C.	AMERICAN BUCKLE & CARTRIDGE CO: West Haven, Conn. U.S.A. (Purchased in 1889).
A. C. CO.	AUSTIN CARTRIDGE CO: Cleveland, Ohio. U.S.A. (1890–1908).
A.L.H.	A. L. HOWARD & CO: New Haven, Conn. U.S.A. (1885).
A.M.C. & CO.	AMERICAN CARTRIDGE & AMMUNITION CO: Hartford, Conn. U.S.A. (1901–1906).
A & N.C.S.L.	ARMY & NAVY CO-OPERATIVE SOCIETY LTD: Westminster, London. U.K. (1909).
A.W.G.	A.W. GAMAGE: Holborn, London. U.K.
AZOT.	BAIKAL: U.S.S.R.
B.P.D.	BOMBRINI PARODI DELFINO: Italy.
B.S.A.	BIRMINGHAM SMALL ARMS. B.S.A. LTD: Birmingham (Midlands). U.K.
C.A.C.	COLONIAL AMMUNITION CO: Auckland, New Zealand.
C.A. CO.	CHICAGO ARMS CO: Chicago, Ill. U.S.A.
C.C.C.	CREEDMORE CARTRIDGE CO: Barberton, Ohio. U.S.A. (1980).
C.C. & T. CO.	CHAMBERLIN CARTRIDGE & TARGET CO: Cleveland, Ohio. U.S.A. (Purchased by the Remington Arms Co, 1930).
C.F.C. CO.	C. F. COOK & CO: (Address not known. Believed to be U.S.A.).
C.I.L.	CANADIAN INDUSTRIES LTD: Brownsburg, Quebec. Canada. (1902 onwards).
D.C. CO.	THE DELAWARE CARTRIDGE CO: Wilmington, Delaware. U.S.A. (1876–1885).
D.C. CO.	DOMINION CARTRIDGE CO: Brownsburg, Quebec. Canada.
E-K.	ELEY-KYNOCH: Witton, Birmingham. U.K. Also, Australia and New Zealand. (1926 onwards).
E.T.L.	EXPLOSIVES TRADES LTD: Witton, Birmingham (Midlands). U.K. (1918–1919).
F.N.	FABRIQUE NATIONALE d'ARMES de GUERRE: Belgium.
GECCO.	GUSTAV GENSHOW & CO: Durlach. Germany.
G-F.	GIVLIO FIOCCHI: Lecco. Italy.
H.S.A.	HAMMERSTOM'S SMALL ARMS: Regina, Saskatchewan. Canada.
H.S.B. & CO.	HIBBARD SPENCER & BARTLETT CO: Chicago, Ill. U.S.A. (Hardware distributors).
H.K.	HAERENS KRUDTVAERK: Denmark.
I.C. CO:	SPORTSMAN'S INTERNATIONAL CARTRIDGE CO: Kansas City. U.S.A. (1914).
I.C.I.	IMPERIAL CHEMICAL INDUSTRIES LTD (METAL DIVISION): U.K. Also, Australia and New-Zealand (Parent firm at Witton. 1926–1970).
I.C.I.A.N.Z.	IMPERIAL CHEMICAL INDUSTRIES of AUSTRALIA & NEW-ZEALAND (1936–1956).

I.M.I.	IMPERIAL METAL INDUSTRIES (AUSTRALIA) LTD: Melbourne, Vic. (1970 onwards).
J.G.R.	J.G.R. GUNSPORT LTD: Toronto, Ontario. Canada.
J.P & S.	J. PAIN & SON: Salisbury, Wilts. U.K.
L.B.	LEON BEAUX: Milan. Italy.
L.B.C.	LEON BEAUX & CO: Milan. Italy.
L.C.M & Co.	LATIMER CLARK MURIHEAD & CO: Millwall, London. U.K. (Founded 1887. Made military ammo. Just possible that some may have been loaded with shot).
M.F.A. CO.	MERIDAN FIREARMS MANUFACTURING CO: Meridan, Conn. U.S.A. (1900).
M.G.M.	MANUFACTURE GENERALE de MUNITIONS: Bouge Les Valence. France.
N. A & A. CO.	THE NATIONAL ARMS & AMMUNITION CO: Birmingham, (Midlands). U.K. (1872–1873).
N.A.C. CO.	NORTH AMERICAN CARTRIDGE CO: Canada.
N.I.	NOBEL INDUSTRIES LTD: Witton, Birmingham (Midlands). U.K. (1919–1921).
N.Y.C.	NEW YORK CARTRIDGE CO: New York. U.S.A.
N.Z.	COLONIAL AMMUNITION CO: Auckland. New-Zealand (1938–1945).
P.C.C.	PETER'S CARTRIDGE CO: Cincinnati, Ohio. U.S.A. (1890 onwards).
R.A.	REMINGTON ARMS CO: Bridgeport, Conn. U.S.A. (1943. On tracer shells used for training air gunners).
REM-UMC	REMINGTON PETERS CO: Bridgeport, Conn. U.S.A. Also Canada.
R.H.A. CO.	ROBINHOOD AMMUNITION CO: Swanton, VT. U.S.A. (Purchased by Remington-UMC Co in 1916).
R.M.C.	RODEN MANUFACTURING CO: Capetown. South Africa.
R.W.S.	RHEINISCHE WESTFALISCHE SPRINGSTOFF: Germany.
S.B. } S & B. } S.B.P. }	SELLIER & BELLOT: Prague. Czechoslovakia.
S.C. CO:	STRONG CARTRIDGE CO: New Haven, Conn. U.S.A.
S.F. } S.F.M. }	SOCIETE FRANCAISE des MUNITIONS de CHASSE et de GUERRE: France.
S.G. Co.	SCHULTZE GUNPOWDER CO: Eyeworth, Hants & Bucklersbury, London. U.K. (Prior to 1914).
S.I.C. CO:	SPORTSMAN'S INTERNATIONAL CARTRIDGE CO: Kansas City. U.S.A.
S.M.I.	SOCIETA METALLURGICA ITALIANA: Italy.
S.O.C. CO.	SOUTHERN CARTRIDGE CO: Houston, Texas. U.S.A.
S.P & A. CO.	SMOKELESS POWDER & AMMUNITION CO. (Address not known. Believed to be U.K.).

S.Q.	WINCHESTER REPEATING ARMS CO: New Haven, Conn. U.S.A. (Used only on certain shells).
SUPER	THE SUPER CARTRIDGE CO: Australia.
S.W.F.	SMITH, WESSON & FIOCCHI: U.S.A. (1973).
S & W.	SMITH & WESSON: U.S.A. (1974. Fiocchi having dropped out).
U.M.C. CO:	UNION METALLIC CARTRIDGE CO: Boston & Bridgeport. U.S.A.
U.S.C. CO.	UNITED STATES CARTRIDGE CO: Lowell, Mass. U.S.A. (Terminated 1929).
V.L & A.	VON LENGERKE & ANTONIE: Chicago, Ill. U.S.A.
V.L & D.	VON LENGERKE & DETMOLD: New York City. U.S.A. (Sporting goods dealer).
W.A.	WESTERN AUTO STORES: U.S.A.
W.C.C. W.C. CO.	WESTERN CARTRIDGE CO: East Alton, Ill. U.S.A.
W.R.A. CO.	WINCHESTER REPEATING ARMS CO: New Haven, Conn. U.S.A. (1884 onwards).
W-W.	WINCHESTER-WESTERN DIV OLIN CORP: New Haven, Conn. U.S.A. Also, Australia & Italy. (1971).

Significant Events

1888
The Union Metallic Cartridge Co. Ltd, of Bridgeport, Conn, U.S.A. combined forces with Remington; Patents were first taken out by The Nobel Explosives Co. Ltd for their 'Ballistite' smokeless powder.

1891
The Arms & Ammunition Manufacturing Co. Ltd., was known to have been in business at 143 Queen Victoria Street, London. Also in Birmingham for the next nine years.

1893
G. Kynoch & Co., Ltd., bought the established gunpowder factory of Messrs Shortridge & Wright at Barnsley.

1894
John Hall introduced his famous 'Cannonite' gunpowder.

1895
Nobel's 'Ballistite' was first marketed in May; Messrs Eley Brothers introduced their 'Pegamoid' Brand cases.

1896
G. Kynoch & Co, Ltd established a new factory at Arklow, Northern Ireland, for the manufacture of cordite.

1897
Kynoch Ltd was formed as a new company. Its main works were The Lion Works at Witton, Birmingham. They also acquired The West Riding Candle Co. Ltd and transferred the whole of this plant to Witton.

1898
The amalgamation of Curtis's & Harvey's and Messrs John Hall & Son; Kynoch & Co, Ltd had a London office in St James's Street.

1901
The Army & Navy Co-operative Society opened new branches in Bombay, Delhi, Calcutta and Karachi.

1902
The Breakable Wad Company Ltd was known to be in existence at Newcastle-upon-Tyne. They were manufacturing wads in Ga's 12, 16 and 20.

1907
The Nobel Explosives Co. Ltd gained complete control of Frederick Joyce & Co. Ltd for their own cartridge case manufacture.

1911
Messrs Eley Brothers acquired the Schultze Gunpowder Co. Ltd.

1918
November of this year saw the formation of a new company. This was the Explosives Trades Ltd. Some cartridges bore the letters E.T.L. on their stampings. Seventeen firms took part in the merger together with their subsidiary companies. The main firms were Eley Brothers; Kynoch & Co. Ltd; Nobel's Explosives Ltd; Messrs Curtis's & Harvey's, bringing with them the Chilworth Gunpowder Co. Ltd; also the E.C. Powder Co. Ltd.

1920
The Ardeer Gunpowder Factory commenced manufacture of Schultze Powder; Explosives Trades Ltd, having lasted as such for two years and now a combination of 40 companies, then changed its name to Nobel Industries Ltd. Some cartridges bore the letters N.I. on their stampings.

1923
The Ardeer Gunpowder Factory commenced manufacture of E.C. Powder.

1926
The Remington Arms Co. made their 'Kleanbore' Cartridge available. All the sporting ammunition of Nobel Industries Ltd was transferred fully to the Kynoch Lion Works at Witton, Birmingham. Nobel Industries Ltd was taken over by the Imperial Chemical Industries Ltd. After this date all cartridges except ·410 and smaller bore had the letters I.C.I. on their stampings.

1932
About this time the Winchester Repeating Arms Co was taken over by the Western Cartridge Co.

1933
E. I. Du Pont de Nemours & Co purchased a 60 per cent interest in Remington.

1934
The Ardeer Gunpowder Factory commenced manufacture of Smokeless Diamond Powder. The merger of Remington & Dupont and the Peters Cartridge Co.

1938
Mortimer & Son Ltd of Edinburgh were bought out.

1947
James Dixon & Sons, Ltd of Sheffield purchased the firm of Geo & J. W. Hawksley of London & Birmingham. Both these firms at one time made powder flasks.

1958
This was the first year of the National Champion Clay Shoots in Australia.

1960
Remington released their plastic shotgun shell on to the market.

1963
Remington introduced a new one-piece plastic wad column giving improved patterns. They also made obsolete their paper-tubed cases.

1966
Winchester started manufacturing shotgun cartridges in Australia. The works were situated in Geelong, Victoria.

1973
During April the first of the new-look cartridges by Eley Ammunition Div. I.M.I. (Kynoch) Ltd were advertised. This was the 70 mm 'Olympic Trap'. The introduction of their new ribbed polyethylene-tubed cases with silver-coloured steel heads.

Index

'Ace' cartridge, 129
   closure on, 129
advertisements, copying for reference, 17
alarm guns, 18–19
   blanks for, 18
ammunition sent by post, 9
arms sales, cartridges at, 10
atmosphere, importance when storing live cartridges, 12
Australian Cartridge Enterprises Limited, 129

Baron, M. le, and electrical gun, 21
barrel (gun), preparing for firing old caps, 9
battery, for electrical gun, 21
beading, in cartridge drawers, 11
beeswax, on paper tubes, 16
Bellford, Mr., and centre-fire cartridge, 20
Biggs, George, 129
brass tubes, pinfire, 18
brasswork:
   cleaning, 13
   cleaning corroded, 15–16
   cleaning on old cartridges, 13
breechloaders, 18
   Schneider and, 20
building a collection, 9

cabinets:
   collections in, 11
   converting, 11
   with drawers, 11
   purpose-built, 11
   strengthening, 11
   wall-mounted, 11
candle wax, use when closing cartridges, 15
caps:
   firing old, 9
   keeping old, 9
   removing live, 9
   replacing fired, 13
card sheeting, in cartridge drawers, 11
cardboard, corrugated, in cartridge drawers, 12
cardboard boxes, collections in, 11
cartridge boxes:
   collecting, 17
   collecting in America, 17
   information from, 17
   using labels from, 11
cartridge code, 25, 29

cartridges:
   Australian, 124ff.
   Australian, post Second World War, 129
   black powder, 9
   cleaning old, 13
   collecting empty, 17
   collecting live, 10
   condition of, 22
   in drawers, 11
   for electrical guns, 21
   estimating age of, 22
   firing old, 9
   firms' names found within, 130
   in glass-fronted cabinets, 11
   information for dating, 17
   information from boxes, 17
   manufacturers/suppliers, 32ff.
   manufacturing, 24
   marketing, 24
   method of loading by hand, 13–14
   method of placing in drawers, 11
   new brands, 9
   obtaining, 10
   one-piece, 129
   origin of pinfire, 18
   plastic, 129
   reading information on, 24
   sources, 10
   as source of information, 17
   storing in alphabetical order, 12
   storing away from children, 12
   storing live, 12
   stripping, 9
   unidentified, 123
   unloading old, 9
   weight of collection, 11
   see also under individual types
cases:
   acquiring modern, 10
   closing plastic, 15
   collecting, 10
   dummy loading plastic, 15
   firing 16-bore in a 12-bore gun, 9
   firing ready-capped, 9
   improving fired modern, 13
catalogues/books, collecting, 17
   use in dating, 22
centre-fire cartridge, 18
   finding early, 9
   invention of true, 20

Lancaster type, 20
charges, removing old, 9
clay shoots:
   cartridges for, 129
   collecting at, 10
cleaning:
   corroded brass, 15–16
   paper tubes, 16
   practising techniques, 16
closing machines, 14
   method of operating hand-, 14
   using a standard type, 15
collectors' clubs, joining, 9
Colonial Ammunition Company Limited, 129
   markings on cartridges, 129
corrosion, treating on brass, 15–16
cottage clearance sales, cartridges at, 10
crimp closures:
   collecting cartridges with, 13
   dummy loads, 13
   flattening, 13
   improving appearance of, 13
   pushing back by hand, 13
   smoothing for unloaded appearance, 13
   using reloading tool on, 13
crimp star closures:
   on Australian cartridges, 129
   improving before storing, 13
   pushing back after filling, 13

dates of manufacture, 22
Daw, George H., and centre-fire cartridges, 20
Daw cartridge, 20
   12- and 16-bore, 21
displayboards, manufacturers', 17
displaying collections, 11–12
drawers:
   storing cartridges in, 11
   strengthening, 11
dummy loading:
   before storing, 13
   paper cases, 13, 15
dummy loads, 13

Eley Brothers, Messrs.:
   100 size cartridge boxes, 17
   and Daw cartridge, 20
   markings on C.A.C. cartridges, 129
   needlefire bullets, 21
   and pinfire cartridges, 18
exchange box, keeping, 10

farmers, as source of cartridges, 10
filing cabinets, for storing cartridges, 11

Fiocchi cases:
   headstamp on, 22
   red paper, 129
firms, information on, 22
flash holes, in centre-fire cartridges, 20
fulminate mixture, in centre-fire cartridges, 20

Ga 12 plastic shells, 129
   colours used, 129
gamekeepers, as source of cartridges, 10
glue, type for use in cartridge drawers, 11
Great Exhibition, Paris, electrical gun shown at, 21
grouse-shooting season, collecting during, 10
gum, non-staining, on old paper tubes, 16
gun firms:
   as source of cartridges, 10
   finding information on, 22
   information from cartridge boxes, 17
gunpowder tins, 17
guns:
   electrical, 21
   preparing for firing old caps, 9
   see also individual types

hand-loading machines, 13–14
headstampings:
   enlarging, 17
   importance in dating, 22
   on late pinfire cartridges, 18
   letter combinations, 132ff.
   method for reproducing, 17
   on Winchester-Western Australian shells, 129
heat, excessive, and live cartridges, 12
Holland & Holland Badminton cartridges, 9

I.C.I.:
   in Australia, 129
   12- and 16-bore pinfire cartridges, 18
I.M.I., in Australia, 129
infra-red, effect on cartridge colours, 11
ironmongers, as source of cartridges, 10

Joyce & Co., Frederick, and pinfire cartridges, 18

Kynoch & Co.:
   100 size cartridge boxes, 17
   and pinfire cartridges, 18

Lancaster, Charles, and centre-fire cartridge, 20
'Light Game' brand, 129

loading tools, 13–16

metal polish, using on brass, 15
Mitchell, Ken, 124
muzzle-loaders, 18

Needham, Mr., and the needle-fire gun, 20
needlefire cartridges, 20
    Eley Brothers' type, 21
needlefire gun:
    development of, 20
    drawbacks to, 20

overshot wads:
    displaying, 17
    keeping for reference, 17

paper tubes:
    cleaning on old cartridges, 13
    colour information, 24
    dampness in, 9
    effect of light on, 11
    filling with sand, 13
    masking before spraying brass, 15
    of pinfire cartridges, 18
    polishing, 16
    protecting when firing caps, 9
    refurbishing, 16
    revitalising old, 16
    swelling in, 9
    types of closure, 14–15
petroleum jelly, using on cleaned brass, 16
pheasant-shooting season, collecting during, 10
Pieper of Liège, and electrical gun, 21
pinfire cartridges, 18, 20–21
    drawbacks to, 18
    finding early, 9
    hand-loading, 14
    manufactured in Britain, 18
    origin of, 18
    transporting, 20
    tube types, 18

plastic sheathing, for protecting pinfire cartridges, 20
polythene, use when closing cartridges, 15
polyurethene lacquer, using on cleaned brass, 15
Pottet, M., and centre-fire cartridge, 20
printing, care of when cleaning, 16

'Quail Load' brand, 129

'Rabbit Load' brand, 129
refurbishing tools, 13–16
reimbursement, offering for cartridges, 10
rolled-top cartridges, finding early, 9
rolled turnover closure, 14

sand, for filling used cartridges, 13
Schneider, François Eugene, and centre-fire cartridge, 20
Schneider, George, patents for breech-loaders, 20
scrapbook, keeping for relevant information, 17
Shawcross, Geoff, 124
shot column, corrosion of, 9
'Special Smokeless' headstamp, 22
storage, indication of poor, 9
stringers, wooden, in drawers, 11
Super Cartridge Company Limited, 129
'Super Express' cartridge, 129

tissue paper, for cleaning paper tubes, 16

wad, removing from ready-capped case, 9
washing-up liquid, for cleaning paper tubes, 16
Wastie, Mr., 11
    method of storing cartridges, 11–12
wet and dry paper, using on corroded brass, 15
Winchester Division, Olin, Australia Limited, 129
    brand names, 129
Winchester-Western, in Australia, 129
wire wool, using on corroded brass, 15